AN INTRODUCTION TO THE ANALYSIS OF ANCIENT TURBIDITE BASINS FROM AN OUTCROP PERSPECTIVE

by

**Emiliano Mutti, Roberto Tinterri, Eduard Remacha*, Nicola Mavilla,
Stefano Angella and Luca Fava**

Dipartimento di Scienze della Terra, Università di Parma,
viale delle Scienze 78, 43100 PARMA, Italy
e-mail: mutti@nemo.unipr.it

*Departament de Geologia, Universita Autonoma de Barcelona
edifici CS, 08193 Bellaterra, BARCELONA, Spain

AAPG Continuing Education Course Note Series #39

Published by
The American Association of Petroleum Geologists
Tulsa, Oklahoma, U. S. A.
Printed in the U. S. A.

Published by the Education Department of
The American Association of Petroleum Geologists
Copyright © 1999 by
The American Association of Petroleum Geologists
All Rights Reserved
Printed in the U.S.A.

ISBN: 0-89181-188-5

AAPG grants permission for a single photocopy of an item from this publication for personal use. Authorization for additional copies of items from this publication for personal or internal use is granted by AAPG provided that the base fee of $15.00 per copy is paid directly to the Copyright Clearance Center, 222 Rosewood Drive, Danvers, Massachusetts 01923 (phone: 978/750-8400). Fees are subject to change. Any form of electronic or digital scanning or other digital transformation of portions of this publication into computer-readable and/or transmittable form for personal or corporate use requires special permission from, and is subject to fee charges by, the AAPG.

Cover from: Proterozoic Turbidite System, Namibia. Note the impressive lateral continuity of turbidite sandstone lobes. Younging direction is from left to right (Mutti et al., 1999).

THE AMERICAN ASSOCIATION OF PETROLEUM GEOLOGISTS (AAPG) DOES NOT ENDORSE OR RECOMMEND ANY PRODUCTS OR SERVICES THAT MAY BE CITED, USED OR DISCUSSED IN AAPG PUBLICATIONS OR IN PRESENTATIONS AT EVENTS ASSOCIATED WITH THE AAPG.

This and other AAPG publications are available from:

The AAPG Bookstore
P.O. Box 979
Tulsa, OK 74101-0979
Telephone: 1-918- 584-2555 or 1-800-364-AAPG (USA)
Fax: 1-918-560-2652 or 1-800- 898-2274 (USA)
www.aapg.org

Geological Society Publishing House
Unit 7, Brassmill Enterprise Centre
Brassmill Lane, Bath, U.K.
BA1 3JN
Tel +44-1225-445046
Fax +44-1225-442836
www.geolsoc.org.uk

Australian Mineral Foundation
AMF Bookshop
63 Conyngham Street
Glenside, South Australia 5065
Australia
Tel. +61-8-8379-0444
Fax +61-8-8379-4634
www.amf.com.au/amf

Affiliated East-West Press Private Ltd.
G-1/16 Ansari Road   Darya Ganj
New Delhi   110 002
India
Tel +91 11 3279113
Fax +91 11 3260538
e-mail: aewp.newdel@aworld.net

An Introduction to the Analysis of Ancient Turbidite Basins from an Outcrop Perspective

Course Text

NOTE

These course notes, co-authored by some of my graduate students involved in deep-water and fluvio-deltaic sedimentation research, summarize four decades of field work by the senior author in many turbidite basins worldwide and particularly, in co-operation with Eduard Remacha during the last 15 years, in the Eocene strata of the south-central Pyrenean foreland basin.

Most of the data and interpretations presented are previously unpublished and therefore, in many respects, the notes constitute a kind of research paper, particularly where facies, processes and cyclicity are concerned. We are aware that many of our conclusions may seem controversial and far from conventional. However, we felt that in recent years there has been little variety in the approach to turbidites and there was a need for fresh air from the area of outcrop studies.

For the sake of brevity, many problems are dealt with in a very cursory manner but the reader can take advantage of the many bibliographic references which will help him to delve deeper into these problems.

The main scheme of these notes is an attempt to consider turbidite sedimentation of ancient orogenic belts as closely related to that of marginal flood-dominated fluvio-deltaic systems, both types of sedimentation being primarily controlled by tectonism and high-frequency climatic variations. Catastrophic floods are thought to represent the fundamental process of this kind of sedimentation which we have provocatively termed "fluvio-turbidite sedimentation". This kind of sedimentation simply suggests that the final depositional zone of highly-efficient, flood-dominated river systems is located in adjacent deep-water basins and is recorded by turbidite accumulations. Of course, we are aware that while catastrophic floods directly control hyperpycnal flows in shallow marine regions, the relations between hyperpycnal flows and deep-water turbidity currents are not that simple and sediment failures and large-scale bed erosion also play a major role here. In particular, we try to emphasize in these notes that real turbidity currents can originate only after a substantial flow acceleration along steep submarine conduits.

Turbidity currents are considered as bipartite currents in which a basal granular layer flows primarily due to inertia conditions and excess pore pressure and is overlain by a turbulent layer which will eventually rework and oudistance the final deposit of the inertia layer in a basinward direction. This scheme strictly conforms to the definition of a turbidity current as given by Kuenen (in Sanders, 1965), thus permitting us to overcome many conceptual and terminology problems discussed in subsequent literature. An extreme example of these problems can be found in a recent paper by Shanmugam (in press), who attempts to dismantle the turbidite paradigm. Paradigms are generally short-lived by nature: nonetheless, turbidites apparently have some life in them yet.

In my experience, it is apparent that courses for petroleum geologists are not infrequently devised to reassure them that existing models are good and that these models have a highly predictive value. Most of these models were actually proposed long ago and would benefit from certain modifications (see Mutti and Normark, 1987). My strong conviction, after many years, is that turbidites are still basically poorly understood and we cannot offer, at present, more than our honest ignorance or little knowledge in this respect. This ignorance is increasingly enhanced by papers which base themselves on terminology problems and disputes, while studies based on basinwide stratigraphic analyses — the only approach by which a geologist can significantly frame facies and facies associations and attempt to develop an understanding of their formative processes — are becoming increasingly rare.

Emiliano Mutti

CONTENTS

1 - Summary 8
2 - Introduction 12
3 - The concept of turbidites: origin and development 14
4 - What is a turbidite? 16
    4.1 - Flowing grain layers and turbulent flows 18
5 - Turbidite facies 20
    5.1 - General 20
    5.2 - The importance of bipartite gravity flows 21
    5.3 - Turbidite facies tracts as related to flow efficiency ... 25
6 - Fan models 28
7 - General characteristics of turbidite systems of thrust-and-fold belts 31
    7.1 - Turbidite systems and their main component elements ... 31
    7.2 - General depositional setting 32
    7.3 - Large-scale submarine erosional features 33
    7.4 - Sandstone lobes and basin-plain deposits 36
    7.5 - Some remarks on the terms "channels" and "lobes" ... 39
8 - Sedimentary cyclicity and sequence stratigraphy of turbidite systems 42
    8.1 - General 42
    8.2 - Floods and turbidity currents 43
    8.3 - Large-scale cyclic stacking patterns 46
    8.4 - Small-scale cyclic stacking patterns 48
References 52
Figures 62

AN INTRODUCTION TO THE ANALYSIS OF ANCIENT TURBIDITE BASINS FROM AN OUTCROP PERSPECTIVE

By

Emiliano Mutti, Roberto Tinterri, Eduard Remacha*, Nicola Mavilla, Stefano Angella and Luca Fava

Dipartimento di Scienze della Terra, Università di Parma,
viale delle Scienze 78, 43100 PARMA, Italy
e-mail : mutti@nemo.ipruniv.it

*Departament de Geologia, Universitat Autonoma de Barcelona, edifici CS, 08193 Bellaterra, BARCELONA, Spain

1 — SUMMARY

- In peripheral foreland basins, turbidite systems are characterized by relatively simple depositional settings. The fundamental expression of these systems are huge accumulations of basinal, cyclically-stacked sandstone lobes with an impressive lateral extent. These accumulations, representing potential reservoirs for hydrocarbons, record the main depositional zone of large-volume turbidity currents originated from hyperpycnal flows emanating from flood-dominated deltaic systems along basin margins. Large-scale submarine erosional features, cut into shelfal and slope sediments, acted as conduits of turbidity currents during their basinward motion. Treating turbidite systems in terms of transfer and depositional zones of turbidity currents may avoid much confusion, both conceptually and in practice. Whatever the depositional model, transfer and depositional zones of turbidity currents will exist in each turbidite basin considered. Like in a fluvial system (in the sense of Schumm, 1977), turbidity currents originate (source zone), flow (transfer zone) and will eventually decelerate to the point that all their sediment load will be deposited (depositional zone). Regardless of their size, plan-view and cross-sectional geometry, and facies types, these deposits record the depositional zone of the currents considered.

- Turbidite sandstone beds appear to be the deposit of bipartite turbidity currents consisting of a basal, high-density and overpressured granular flow overlain by a low-density, fully turbulent flow. These two flows are the basic components of a turbidity current whose sediment load includes coarse-grained sediment. This concept is supported by both theoretical and experimental work (e.g., Sanders 1965; Ravenne and Beghin, 1983; Norem et al., 1990). Conglomerate, pebbly-sandstone and relatively coarse-grained sandstone facies are the typical deposit of granular flows. The genuine Bouma sequence, as intended in these notes, forms farther basinward and is primarily the deposit of relatively low-density and fully turbulent flows which have outdistanced their parental inertia flows in basinal regions.

- A basal, overpressured granular flow, mainly driven by inertial forces and excess pore pressures, is needed to form laterally extensive beds of relatively coarse-grained sediment, i.e. the turbidite reservoir facies. This kind of flow has been described and termed in many different ways (e.g., high-density turbidity current, modified grain flow, debris flow, hyperconcentrated flow, flow-slide), thus generating considerable confusion among sedimentologists and stratigraphers.

- Coarse-grained granular flows, probably originated directly from flood-generated subaerial flowslides and their bulking through erosion and sediment failures in delta-front regions, are funneled in relatively steep submarine conduits along which they progressively accelerate until reaching a phase of catastrophic bed erosion. As a consequence, a large amount of fine-grained sediment is incorporated within the upper, turbulent part of each flow, thus increasing its density, thickness and velocity. At this point a fully-developed, bipartite turbidity current is formed.

- All other things being equal, the processes controlling sediment transport and deposition within a turbidity current during its basinward motion are essentially related to the ability of the basal granular flow to maintain its excess pore pressure and to the amount of turbulent energy developed in the upper and more dilute part of the flow. The first factor controls the runout distance of the granular flow and therefore the lateral extent of coarse-grained facies. The second factor controls how much fine-grained sediment (fine sand and mud) can be resuspended from the basal granular layer and transported farther basinward, and how much traction the turbulent flow can exert on the residual coarser-grained sediment of the granular layer. These two factors combine in determining the degree of efficiency of a turbidity current, i.e. its ability to carry its sediment load in a basinward direction and segregate the original grain populations of this load into distinct and relatively well-sorted facies types with distance. Flow efficiency is therefore of fundamental importance in controlling geometry and reservoir characteristics of turbidite sandstone beds.

- Highly efficient turbidity currents dominate transport and deposition in large and elongate foreland basins. Facies tracts reconstructed from these basin fills through careful stratigraphic correlations permit the recognition of the many different processes (erosion, water escape, bypass, reworking, deposition) that characterize a turbidity current during its basinward motion. Facies and processes can thus be framed into models of fairly good predictive value.

- Particularly during their earlier stages of development, foreland basins have apparently a relatively simple topography. Turbidity currents accelerate when moving along their steeply sloping conduits and decelerate at the exit of these conduits due to the lateral spreading of the flow. After moving across substantially flat basin-proximal regions (lobe region), where they progressively outdistance their parental granular flows, turbulent flows may reach sufficient thickness to experience reflection from bounding slopes and ponding in terminal basin-plain regions. Structurally complex settings require that facies distribution pattern be considered only on a case-by-case basis.

- Turbidity currents may be forced to deposit much of their sediment load without developing a significant segregation of their load if: (1) local topographic obstacles decelerate the currents; (2) parental granular flows cannot reach the catastrophically erosive phase either because they are too thin (small volume) or move on insufficiently steep slopes; (3) parental granular flows cannot maintain their excess pore pressure long enough because of their low mud content. In all these cases, turbidity currents will be poorly efficient, generating facies tracts composed of a limited number of facies, each characterized by relatively poor sorting and, therefore, poor reservoir quality.

- Sequence stratigraphy and sedimentary cyclicity of turbidite basin fills has to be treated with great caution. As amply discussed in these notes, turbidite systems of foreland basins are invariably associated with tectonically-induced angular unconformities along basin margins, suggesting that tectonic uplift must play a major role in producing large sediment availability and high-gradient fluvial systems. As suggested by Milliman and Syvitski (1992), Mulder and Syvitski (1995) and Mutti et al. (1996), sediment flux to the sea is dramatically increased in settings of this type through flooding. Sedimentary cyclicity of turbidite systems of foreland basins strongly suggests that climatically-induced high-frequency ciclicity (Milankowich cyclicity) punctuated low-frequency cycles governed by tectonic uplift, subsidence and denudation.

2 — INTRODUCTION

The vast majority of exposed ancient turbidite systems occur in thrust-and-fold belts where they constitute the fill of both peripheral foreland and associated wedge-top basins (Fig. 1). Since the work of Kuenen and Migliorini (1950), most of our knowledge of turbidite sedimentation is still based on the study of these systems and on the models derived from them.

However, it has become clear with time that turbidites, both recent and ancient, occur in many other types of basins which substantially differ from those of thrust-and-fold belts in terms of tectonic setting, physiography and feeder systems (e.g., Stow et al., 1985; Mutti and Normark, 1987; Pickering et al., 1989; Weimer and Link, 1991). In particular, large accumulations of deep-water sandstones, mainly turbiditic in origin and associated with large river systems, occur in offshore basins of divergent margins (Fig. 1) representing targets of paramount importance for present and future oil exploration (Weimer and Link, 1991). Questions like how and to what extent depositional models derived from the study of ancient turbidite systems of thrust-and-fold basins can be used to describe these systems and to predict their sand distribution patterns present a challenge for future sedimentological research. Most ancient turbidites of orogenic belts appear to have formed in narrow and elongate basins that experienced little, if any, reworking by oceanic bottom currents (Mutti and Normark, 1987, 1991). These tectonically active basins, typically developed on continental crust, are thus ideal to study turbidite sedimentation without any substantial interaction with other deep-marine processes. Conversely, bottom currents and their deposits (contourites) are common in most oceanic basins, particularly on divergent margins (e.g., Pickering et al., 1989; Faugères & Stow, 1993; Shanmugam et al., 1993; Viana et al., 1998; with references therein). To what extent bottom currents can interfere with and modify the depositional patterns of turbidite systems developed on these oceanic basins is poorly understood and will also require a considerable research effort in future years.

The main purpose of the notes is to provide criteria that can be used for the description and interpretation of ancient turbidite systems exposed in thrust-and-fold belts. These systems are typically expressed by thick and laterally continuous sandstone lobes which grade peripherally into finer-grained basin-plain facies. Classic examples of these systems are exposed in the south-central Pyrenees and the northern Apennines, forming the fill of elongate basins stretched parallel to advancing thrust fronts.

These turbidites were deposited by large-volume flows which were able to travel considerable distances across the basin floor and whose conduits are represented by basin-margin large-scale erosional features. The final depositive stage of these flows is represented by waning and depletive turbidity currents which are commonly reflected by basin margins and may undergo ponding in terminal basin plain regions. These notes focus in particular on the coarse-grained facies deposited by submarine gravity flows before reaching this final depositive stage during which turbidite sedimentation is mainly restricted to fine and very-fine sand and mud.

In our experience, it is quite clear that significant studies of turbidite systems have to be based on careful facies analysis framed within high-resolution stratigraphic correlation patterns covering significant portions of a basin fill. Facies analysis and detailed sedimentological descriptions carried out on small portions of a basin fill or on isolated outcrops and sections can be highly misleading. In addition, growing evidence from both recent and ancient settings indicates that our understanding of turbidite depositional systems can be greatly improved through the study of the associated fluvio-deltaic systems which are preserved along the margins of turbidite basins.

3 — THE CONCEPT OF TURBIDITES: ORIGIN AND DEVELOPMENT

Apart from a very short contribution by Migliorini (1943) on the deposits of submarine density currents — a milestone paper which was essentially ignored in subsequent literature probably because it was written in Italian -, the concept of turbidite sedimentation originated from the fundamental paper by Kuenen and Migliorini (1950) on "graded sandstone beds" resedimented by turbidity (density) currents. The paper was based on the ideas of Ph. H. Kuenen on turbidity currents developed through laboratory experiments and marine geologic observations and on C.I. Migliorini's work on the Paleogene strata of the Scaglia and Macigno formations in the northern Apennines.

These new concepts led to a series of papers by Ph. H. Kuenen and his students and co-workers on turbidite strata of many ancient basin-fills during the 50's, particularly in the northern Apennines, Carpathians, Alps and California (see summary and references in Kuenen, 1964). Much of this work highlighted the deep-water nature of turbidites, their main sedimentary features, particularly grading and sole markings, and the importance of resedimentation processes through turbidity currents in tectonically active basins. The term "turbidites" was introduced by Kuenen (1957, p.231) following a verbal suggestion by C.P.M. Frijlinck.

By 1964, the number of papers dealing with turbidites had grown to 650 (see Kuenen and Humbert, 1964). It would be impossible to list the many thousands of papers that have discussed the different aspects of turbidite and turbidite-related sedimentation in both modern and ancient strata between 1964 and the end of this millenium. Needless to say, the concept has been among the greatest breakthroughs in sedimentary geology (see Friedman and Sanders, 1997).

Today, the concept of turbidites is accepted by sedimentary geologists and generally associated with deep-water sedimentation. Although turbidites are under-represented in modern basins and turbidity currents are inherently difficult to observe directly, some examples of modern turbidite deposits have been studied in detail. The best known example is certainly that of the Grand Banks on the Laurentian fan, which was triggered by an earthquake in 1929 (see summary in Normark and Piper, 1991). Other well known examples include turbidites deposited in fjord settings and originated from delta-front instability processes directly or indirectly related to fluvial floods (e.g., Prior and Bornhold, 1989; Zeng et al., 1991).

The problem of the origin and ignition of turbidity currents has been amply discussed by Normark and Piper (1991). The authors conclude that earthquakes, sediment failures, storms and rivers in flood are the most common triggering mechanisms that generate turbidity currents. More recently, growing and convincing evidence from both modern and ancient basins suggests that hyperpycnal flows generated by rivers in flood play a major role in the origin of turbidity currents in marine basins, implying a close genetic relation between fluvio-deltaic and turbidite sedimentation (Milliman and Syvitski, 1992; Mulder and Syvitski, 1995; Mutti et al., 1996).

Following Kuenen's (1950) pioneering work, many attempts have been made to reproduce turbidity currents in laboratory experiments (e.g., Middleton, 1966a, 1966b, 1967; Luthi, 1981; Postma et al., 1988; Ravenne and Beghin, 1983; Parker, 1982; Gladstone et al., 1998). The results of these experiments have shed considerable light on many hydrodynamic problems related to these currents but, because of scale problems, have generally failed to reproduce deposits comparable to those that can be studied in outcrops. More recently, numerical modeling based on laboratory experiments, monitoring of modern turbidity currents in ocean basins and some field data have also been used to develop a better understanding of turbidity currents and their deposits (e.g., Kneller, 1995; Zeng and Lowe, 1997a, 1997b). Most commonly, however, modeling has been restricted to sandy turbidites transported and deposited by turbulent currents, thus overlooking coarse-grained deposits that are a fundamental component of most turbidite successions (see later).

Most of the above experimental work has focused on surge-type turbidity currents, i.e. catastrophic flows of relatively small volume and short duration. This concept has long been implicit in most literature. Today, it is obvious that many turbidity currents can last hours and days (e.g., Normark and Piper, 1991; Mulder et al., 1998), thus suggesting that these flows are maintained by a continuous sediment supply from behind, most commonly by hyperpycnal flows. In a paper mainly concerned with the deposition of thick massive turbidite sands, Kneller and Branney (1995) have contrasted this type of turbidity current, which they termed "sustained turbidity current", with surge-type currents and emphasized the differences in the style of deposition of the two types of flow (see later).

4 — WHAT IS A TURBIDITE?

To most geologists a turbidite is a graded bed consisting of a sandstone/mudstone couplet which has been deposited by a turbidity current and is commonly overlain by a hemipelagic mudstone containing deep-water fossil assemblages. The Bouma sequence typifies this ideal kind of deposit (Fig. 2a).

Bouma (1962) developed his classical model essentially from observations made in upper Paleogene and Neogene turbidite strata of the western Alps (Annot Sandstone) and the northern Apennines (Macigno and Marnoso-arenacea formations). Although commonly restricted to the internal sequence of depositional divisions observed within an individual turbidite bed, the Bouma model actually includes the fundamental concept of "depositional cone" (Fig. 2b), implying the depletive character of turbidity currents, i.e. their deceleration in space away from their origin. On the basis of this concept, Parea (1965) and Walker (1967) developed their models of proximal vs. distal turbidite deposition — an approach which in many respects still has considerable potential for future research.

Bouma (1962) interpreted his sequence as the deposit of a typical turbidity current without discussing hydrodynamic processes in detail; Harms and Fahnestock (1965) and Walker (1967) reinterpreted the sequence in terms of laboratory flow regime postulating an upper flow regime for the basal *"a"* division (Fig. 3). In a very influential paper, Middleton and Hampton (1973) suggested that the entire Bouma sequence was the deposit of a fully turbulent turbidity current and envisaged the *"a"* division as the result of very rapid deposition from suspension preventing the formation of traction structures.

Subsequent work has shown that the Bouma sequence is an oversimplification of the many possible expressions of turbidite sandstone beds and, ironically, the basic question remains as to what a turbidite bed is. Is this bed the deposit of a fully turbulent density current that moves downslope because of its excess density or is it rather the deposit of a more complex flow whose final state is that of a fully turbulent turbidity current? In other words, should we consider as turbidites only those beds that can be clearly interpreted as transported and deposited by a turbidity current, i.e. a turbulent flow where the particles are maintained within the flow by turbulence, or should we extend the definition of turbidites to include all those beds which appear to have formed through genetically related gravity flows, the last of which is a waning and depleting turbidity current?

The problem is a long-standing one that originated with the concept of "fluxoturbidite" of Dzulynski et al. (1959) and particularly with that of the "flowing grain layer" of Sanders (1965), both derived from beds or portions of them implying transport and deposition by types of flow in some ways different from fully turbulent turbidity currents. More recently, Shanmugam et al. (1994) and Shanmugam (1996, in press) have raised the same problem, suggesting that most thick and coarse-grained sandstone beds observed in deep-water successions are deposits of sandy debris flows, not turbidity currents.

At this point one might also wonder what Kuenen and Migliorini actually thought was a "classic" turbidity or density current such as was implied in their original paper. There seems to be little doubt that their graded resedimented sandstone beds also contain basal coarse-grained divisions deposited by the lower and denser part of the flow and that the authors included these divisions in their turbidite concept. These coarse-grained, commonly structureless, divisions and associated pebbly sandstones and conglomerates, are those which have raised most problems in recent years concerning their formative processes and the relations of these processes to fully turbulent turbidity currents. Although semantics may have played a role in adding further confusion, the problem does exist and entails all those facies and processes that refer to debris flows, both cohesive and cohesionless, and high-density, gravelly and sandy, turbidity currents, i.e. processes where high sediment concentration usually damps turbulence and hinders settling.

4.1 — Flowing grain layers and turbulent flows

In an extremely important and generally overlooked paper, Sanders (1965) first interpreted the Bouma sequence as the deposit of two different kinds of flow: a basal and faster moving inertia flow (or flowing grain layer) and an overlying turbulent flow to which he restricted the term of turbidity current (Fig. 4). In the basal flow, coarser grains would move essentially under "bed-load" conditions and the mixture of sediment and water would be impelled by the shear stress imparted from the overlying turbulent flow where finer grains are individually kept in suspension by turbulence. As a result, the "*a*" division of the Bouma sequence would be the *"en masse"* deposit of an inertia flow which underwent freezing because of internal friction, whereas the overlying "*b*" through "*e*" divisions would represent the normal deposit of a turbulent suspension through a traction-plus-fallout process. The "*b*" through "*e*" divisions would therefore be the real turbidite — a corollary following the assumption that a turbidity current is a fully turbulent suspension.

In connection with the above interpretation Ph.H. Kuenen made the following comment (in Sanders, 1965, p.217): *"Field geologists and marine geologists need stable terms that do not change with every development of insight in mechanisms of transportation. But Sanders wishes to restrict the term turbidite henceforth to the upper parts of the beds designated by that term in present usage, because he claims a non-turbulent mode of transport for the sandy part. If he is shown to be wrong — or partially wrong as I believe — then the definition will again have to be changed. Such a procedure will cause great confusion and frustration. It is much simpler to leave out of the definition of turbidity current any reference to hydrodynamic mechanisms. I suggest: "A turbidity current is a current flowing in consequence of the load of sediment it is carrying and which gives it excess density". This leaves the controversial mode of suspension out of account (turbulence, impact, inertia flow, etc.)."*

This comment is fundamental in many respects. Ph.H. Kuenen was certainly aware of the possibility of other transport mechanisms associated with fully turbulent suspensions (e.g., the "fluxoturbidite" concept of Dzulynski et al., 1959), but he urged sedimentologists to maintain the term "turbidites" since, whatever the mechanisms that sustain the particles within the flow, turbidites are certainly produced by submarine density currents. The deposits of the classic submarine sediment gravity flows of Middleton and Hampton (1973), i.e. a spectrum of density flows including debris flows, grain flows, fluidized sediment flows, and turbidity currents (Fig. 5), fit such a broad definition of turbidites. This concept should solve many terminology problems and long-standing disputes and, more importantly, re-establish a stable nomenclature for geologists.

5 — TURBIDITE FACIES

5.1 — General

Turbidite facies and their inferred processes have been dealt with in many papers among which those by Mutti and Ricci Lucchi (1972, 1975), Walker and Mutti (1973), Walker (1978), Mutti (1979, 1992), Nardin et al. (1979), Lowe (1982), and Pickering et al. (1986, 1989) are the most widely accepted in subsequent literature.

These facies schemes differ from each other in many ways. Some schemes emphasize the objective recognition of facies characteristics, others emphasize the genetic relationships between facies and processes, while others have the tendency to combine descriptive and genetic approaches. The reader is referred to Mutti (1992) for a more extensive discussion.

The genetic approach is process-oriented, implying a continuum of genetically related processes and facies the last expression of which is the deposit of depletive and waning low-density turbidity currents. Following Kuenen's broad definition of turbidites (see above), the approach also implies that the continuum does not only define those beds which can be interpreted as the deposit of a turbulent turbidity current, but also those beds or portions of beds which can be interpreted as the deposit of other kinds of processes which — regardless of their inferred characteristics — are high-density currents most commonly moving under more dilute and fully turbulent flows.

In these notes we will follow the process-oriented "facies tract" concept as discussed by Mutti (1992), that is the downcurrent association of different but genetically related types of facies which are time-equivalent and, within each system considered, record the downstream evolution of sediment gravity flows. An ideal facies tract is recorded by an individual bed which is the product of a single sediment gravity flow undergoing transformations during its basinward motion. More commonly, a facies tract has to be established for groups of beds which are accurately correlated over significant distances. The concept of turbidite facies can be applied to a specific depositional division of individual turbidite beds as well as to the dominant type of division observed in a group of time-equivalent beds. Viewed in this way, a turbidite facies represents the deposit of a sediment gravity flow at a specific location along the path of the flow. The approach implies that, within each system considered, turbidite facies tracts be obviously established on the basis of detailed stratal correlation patterns.

Kneller (1995) and Kneller and Branney (1995) have recently emphasized how submarine topography may dramatically affect local turbidite facies distribution patterns and have therefore highlighted the importance of spatial changes in flow velocity (non-uniform flows). Figure 6 depicts the very important relationships between spatial and temporal accelerations (both positive and negative) that can be experienced by turbidity currents during their motion and the idealized types of beds that can be produced by the two kinds of acceleration. These concepts need to be carefully considered when working in detail on the relationships between local basin topography and sand distribution patterns. However, for the more general purposes of these notes, it is fair to state that, in each turbidite system considered, facies distribution patterns indicate an overall downcurrent decrease in grain-size implying a deceleration of gravity flows in time and space in the same direction.

5.2 — The importance of bipartite gravity flows

Following the concepts discussed in previous sections, we include in our turbidite facies all those sediments deposited by sediment gravity flows, ranging from mud-supported conglomerates deposited by cohesive debris flows to graded mudstones deposited by very dilute turbidity currents. The process-oriented and predictive framework within which these facies can be considered is shown in the diagrams of Fig. 7a, 7b, 7c. Facies types (Fig. 8) are essentially those of the classification proposed by Mutti (1992).

We interpret turbidite facies associations containing a substantial proportion of conglomerates and pebbly sandstones as the deposit of submarine granular flows directly fed by heavily-loaded hyperpycnal flows emanating from coarse-grained delta systems. In our interpretation, these flows represent the downstream evolution of flood-induced flowslides originated through liquefaction in fluvial drainage basins and accelerated along steep and narrow fluvial valleys. Once they have reached the sea and become separated from their more dilute portions (hypopycnal flow), these flows are funneled into submarine conduits as submarine granular flows (Fig. 9).

Submarine granular flows soon become bipartite currents. These bipartite currents include a faster-moving, basal and denser layer (granular or inertia flow) where turbulence is damped by high sediment concentration and a turbulent suspension above. Coarse-grained sediment is restricted to the basal flow in which it is sustained by the complex interaction of high sediment concentration, buoyancy, fluid escape generated by excess pore pressure, fluid strength and dispersive pressure; fine-grained sediment tends to be mainly trasported as suspended load in the upper and more dilute part of the flow.

Although expressed in different terms, this bipartition is essentially what was originally suggested by Sanders (1965), implicitly admitted by Middleton and Hampton (1973, p. 30), and more recently re-emphasized by Ravenne and Beghin (1983) and Norem et al. (1990) on the basis of experimental work and numerical modeling respectively. Based on our experience, these flows represent the most common type of sediment transport and deposition of turbidity currents intended in the broad sense discussed above.

The concept of basal granular layer coincides with that of "high-density turbidity current" (cf. Lowe, 1982), the latter term being very popular among many sedimentologists and probably the easiest to use for general purposes of communication. Following Kuenen's suggestion (see above), the basal granular layer is considered in these notes as strictly part of a turbidity current carrying relatively coarse-grained sediment. A growing body of experimental work suggests that this kind of flow — most commonly referred to as granular flow, flowslide or debris flow — is primarily driven by inertia forces under conditions of high excess pore pressure (e.g., Norem et al., 1990; Inverson, 1997; Marr et al., 1997; Gee et al., 1999). Field evidence (Labaume et al., 1987) and laboratory experiments (Mohrig et al., 1998) show that hydroplaning may occur in the early stages of motion of the flow.

Of course, we are aware that many turbidite successions are dominated by fine-grained facies, suggesting deposition from low-density turbidity currents most of which probably originated directly as turbulent flows. This kind of turbidite sedimention is not dealt with in these notes.

As shown in Fig. 7, during the downslope motion of the basal granular layer, coarser grains outdistance finer grains; as a result, coarser grains collect at the head of the flow giving way to a negative horizontal grain-size gradient in a downcurrent direction (Sanders, 1965; with references therein). When the frontal part of the flow freezes because of internal friction, finer grains which were being transported behind overtake and step over the newly formed deposit and keep moving downcurrent producing another horizontal grain-size gradient within the flow. The process will continue until the basal dense layer is entirely deposited through a series of sedimentation waves each characterized by progressively finer-grained sediment in a downcurrent direction (cf. Lowe, 1982; Mutti, 1992). Successive and progressively finer-grained sedimentation waves may produce substantial sediment bypass along the path of each flow; this process is recorded by distinct breaks in grain size within the deposit. As pointed out by Norem et al. (1990), long runout distances of the basal granular layer (their "submarine flowslide") can only be explained by assuming that a substantial excess pore pressure was present within the flow and maintained by a velocity, i.e. water escape takes place slowly compared to flow velocity.

The overlying turbulent flow thickens downcurrent because of progressive mixing with ambient fluid (Fig. 9). This implies that during its basinward motion the energy of the flow is progressively converted into turbulence at flow boundaries. The upper turbulent flow will eventually outdistance the deposit of the basal granular layer and carry its suspended sediment farther basinward. Briefly, the general process that can be envisaged is one through which the granular layer flows due to inertia forces and progressively increasing shear stress imparted from the overlying turbulent flow, whereas the latter moves as an "ideal" turbidity current in which the excess density is due to its sediment load kept in suspension by turbulence.

The motion of bipartite flows undergoes an important transformation (in the sense of Fisher, 1983) in a downcurrent direction whereby the basal granular layer will undergo progressive elutriation of finer particles because of (1) horizontal outdistancing of the coarser particles with development of a negative grain-size gradient, (2) incorporation of the finer particles into the overlying turbulent flow at the head of the basal layer and along the boundary zone between the two flows, and (3) the upward movement of the finer particles driven by excess pore pressure (water escape). As a result, excess pore pressure within the basal granular layer will be suppressed and the flow forced to come to rest (Fig. 10).

As long as there is substantial excess pore pressure, the boundary between the basal granular layer and the overlying turbulent flow is likely to consist of a relatively thick and highly concentrated fluidized bed maintained by water escape. This bed prevents the formation of a rheological surface along which the overlying flow can rework the top of the granular layer through traction. Traction and reworking may start only when water escape comes to an end. It will be noted that the importance of this dewatering bed has also been postulated by Kneller and Branney (1995), though in the framework of gradual vertical aggradation of sand beneath a sustained steady or quasi-steady turbidity current. In our scheme, the dewatering layer is produced by an upward motion of finer particles contained in the basal granular layer (elutriation); in the Kneller and Branney scheme the same type of layer is produced by the high rate of sedimentation from above.

5.3 — Turbidite facies tracts as related to flow efficiency

Basically, we identify four major turbidite facies groups which are defined primarily by their texture (Fig. 7). The grade classes are the following: (A) boulder- to small pebble-sized clasts, (B) small pebbles to coarse sand, (C) medium to fine sand, and (D) fine sand to mud. It will be noted that these grain-size populations are the same as those of the facies schemes of Lowe (1982) and Mutti (1992).

The above grain populations tend to be transported and deposited by turbidity currents as naturally distinct entities, thus forming similarly distinct facies groups. The first two populations invariably move within a basal granular layer; the third population first moves within the granular layer but can be progressively incorporated as suspended load into the overlying turbulent flow; the fourth population moves preferentially as suspended load within the turbulent flow.

The facies types shown in Fig. 7 are arranged in facies tracts each of which records a different degree of flow efficiency. The concept of flow efficiency has been discussed in many papers (e.g., Mutti and Johns, 1978; Mutti, 1979, 1992; Pickering et al., 1989; Allen and Allen, 1990; Mutti et al., 1994; Richards et al., 1998) with reference to the ability of the flow to carry its sediment load basinward. In these notes, we expand the concept to incorporate also the ability of the flow to segregate its grain populations into distinct facies types with distance. Very highly efficient flows will fully segregate the grain populations contained within the parental flow with distance, thus producing relatively well-sorted facies types. Conversely, very poorly efficient flows will only partly segregate their different grain populations, thus producing a more limited number of facies types characterized by poor textural sorting.

With a bipartite current, efficiency has to be considered separately for the two component flows. All other things being equal, the efficiency of the basal granular layer – i.e. its ability to transport grain populations A and B over considerable distances – is essentially related to the velocity gradient that controls the water escape rate (see above) and the amount of fines contained in the parental flow since fines hinder water escape. The efficiency of the turbulent flow depends primarily on the amount of turbulent energy generated at flow boundaries and, therefore, on the amount of fines (grain populations C and D) that the flow can incorporate from the basal granular layer and bed erosion taking place at the head of the basal flow.

Fig. 7a shows the deposits and inferred processes of an ideal very highly efficient flow which can entirely segregate its coarser-grained particles into F3 and F6 deposits. Finer-grained sediment has been completely resuspended into the turbulent flow and thus carried at relatively long distance farther basinward.

Fig. 7b depicts the facies tract most commonly observed in many ancient turbidite basins. The efficiency is still high but only the coarsest particles of the sediment load have been effectively segregated (F3 deposit). The grain-size population B is largely deposited by a granular flow, experiencing high excess pore pressure, in essentially structureless, poorly-sorted beds (F5 deposit). Only the most downcurrent deposit of this grain-size population has undergone elutriation of the finer particles and subsequently reworked by the overlying turbulent flow into very distinctive bedforms. These include both plane bed and 3-D megaripples sometimes capped by small-scale, coarse-grained ripples. These sediments are collectively referred to as F6 deposits (Mutti, 1992). Cross stratification related to the downcurrent migration of megaripples is the most characteristic feature. Typically, these megaripples have a height of about 17 cm and wave lengths up to three meters. Features of this kind strongly indicate reworking from long-lived and large-volume turbulent flows.

Farther downcurrent, the most common type of bed is an F7 deposit characterized by mm-thick traction carpets of coarse sand alternating with medium and fine sand. These alternations are interpreted herein as the product of a mixed type of sedimentation in which traction of coarse sand at the base of the flow takes place concomittantly with fallout of medium and fine sand from the overlying suspension within an overall aggradational process. Coarse sand is derived from the reworking of the deposit of the granular flow located more upcurrent.

Farther downcurrent, F8 deposits are characteristically composed of predominantly fine-grained sand and lack tractional features. The most plausible interpretation of this kind of deposit is still that offered by Middleton and Hampton (1973). These authors suggested that such beds would result from high rates of sediment fallout from an overlying suspension, thus preventing the formation of traction features and causing liquefaction ("quick bed") because of excess pore pressure (see also Kneller and Branney, 1995). In these notes we suggest that this relatively fine-grained and structureless deposit should be considered the genuine "a" division of the Bouma sequence — a division poorly described in the original definition (Bouma, 1962) and consequently the matter of many controversies.

The most distal facies of this tract is represented by the classic base-missing Bouma sequences, i.e. fine sand and mud deposited through traction-plus-fallout and fall-out processes during the final depletive and waning stages of a turbidity current (F9 deposits). The diagram of Fig. 12b depicts these deposits as being related to a strictly unidirectional flow. In reality, in most thrust-and-fold belt basins this last stage of deposition of turbidity currents is commonly affected by flow reflection and ponding (e.g. Pickering and Hiscott, 1985; Remacha et al., 1998).

The deposits of highly-efficient flows clearly indicate that these flows were generally of large volume and long duration and had considerable amounts of fines in suspension. Stated in other words, these flows were sustained turbidity currents (Fig. 11b). The most plausible origin of such flows are long-lived hyperpycnal flows, though very large slumps may also produce this kind of flows (e.g., the Grand Banks turbidite of 1929; see above).

Fig. 7c shows an attached facies tract formed by a poorly-efficient flow. These flows are generated by the failure of limited volumes of sediment or by small-volume and short-duration floods, thus fitting the concept of surge-type flows quite well (Fig. 11a). Poorly efficient flows cannot effectively segregate their different grain-size populations and the importance of their upper turbulent flow is quite reduced. As a result these sediments are generally poorly sorted and are characterized by a considerably smaller areal extent than those deposited by highly-efficient flows. Typically, the distal and finer-grained deposits of these flows (F9b deposits) are poorly developed, suggesting that the original flows did not contain substantial amounts of fines or the flow did not have enough energy to erode substantial amounts of mud from the bed. The F9b facies shown in Fig. 12a denotes F9-like sediments where traction-plus-fallout structures are poorly devloped because of high-rates of flow deceleration.

The facies tracts described above will be further discussed in the following sections dealing with the elements of turbidite depositional systems.

6 — FAN MODELS

In the early 70's, Normark (1970), Mutti and Ghibaudo (1972) and Mutti and Ricci Lucchi (1972) introduced their popular fan models which attempted to interpret turbidite sedimentation within the framework of deep-sea fan depositional systems.

Normark (1970) developed a model of fan growth based on the study of small modern fans from the California borderland basins and emphasized the importance of depositional bulges or suprafans developed at the terminus of fan valleys (Fig. 13). Each suprafan is described as a convex-upward depositional feature with shallow and ephemeral channels in its proximal sector passing downfan into progressively smoother zones characterized by fine-grained sediments.

Mutti and Ghibaudo (1972) and Mutti and Ricci Lucchi (1972) proposed a model for ancient turbidite systems in which for the first time turbidite facies associations were interpreted in terms of specific deep-sea fan environments. Mutti and Ghibaudo (1972) emphasized the strong depositional similarities between fluvial-dominated deltas and deep-sea fans, suggesting a direct comparison between deltaic channels and mouth bars and turbidite channels and lobes respectively (Fig. 14). Mutti and Ricci Lucchi (1972) offered a more comprehensive model where turbidite facies associations were interpreted in terms of slope, fan and basin plain environments and specific facies associations were interpreted as diagnostic of inner, middle and outer fan sub-environments (Fig. 15). Mutti and Ricci Lucchi (1972, 1974) also stressed the overall progradational character of many ancient submarine fan systems and emphasized the thinning- and fining-upward character of channel-fill sequences contrasting with the thickening- and coarsening-upward character of turbidite sandstone lobes.

An attempt to combine the models of Normark (1970) and Mutti and Ricci Lucchi (1972) was subsequently made by Walker (1978). This model also became quite popular among stratigraphers and sedimentologists. Later, Chan and Dott (1983) and Heller and Dickinson (1985) proposed a ramp model for turbidite systems directly fed from multiple deltaic distributary channels, contrasting with the point-sourced canyon-fed systems of the classic fan model.

With more or less substantial modifications, fan models were widely used (and in some case are still used) to describe and interpret a great number of both ancient and modern turbidite systems for many years. This led to a situation of great confusion summarized by Normark et al. (1993) who wrote that "it seemed that the number of fan models began to approach the number of turbidite systems that had been studied". Despite this general tendency to use fan models and generate new ones (mostly as derivatives of the original ones), growing evidence from outcrop and marine geologic studies, as well as from the interpretation of commercial seismic-reflection profiles, showed the increasing difficulties in using these models to describe and interpret the great variety of both ancient and modern turbidite systems that occur in the stratigraphic record in a similarly great variety of tectonic and physiographic settings. By the early 80's, the scientific community became aware of these problems which were discussed at the first COMFAN (COMmittee on FANs) meeting in 1982 (see summaries in Normark et al., 1983/84, and Bouma et al., 1985).

Turbidite systems, both modern and ancient, can actually range from very large to very small, be sand-rich, mud-rich or mixed, and differ considerably from each other in terms of internal architecture, plan-view and cross-sectional geometry, and facies distribution patterns. Clearly, no general unifying models can satisfactorily describe and explain this complex family of depositional systems. This complexity and the great variety of its expressions in both modern and ancient basins have been discussed in many papers and summarized in text books and special volumes (e.g., Bouma et al., 1985; Pickering et al., 1989; Weimer and Link, 1991; Weimer et al., 1994). More recently, Reading and Richards (1994) and Richards et al. (1998) have provided further insight into the problem. The most recent attempt to develop a model for slope and base-of-slope sedimentation has been offered by Shanmugam (1999, his Fig. 33) in an unrealistic association of slumps, plastic flows, debris flows (both muddy and sandy), contourites, and turbidites, the latter being of very limited relevance. The model considers both channelized and non-channelized systems and presumably refers to a divergent margin setting.

Sequence-stratigraphic concepts (Mitchum, 1985; Mutti, 1985; Vail, 1987; Posamentier et al., 1988; Posamentier et al., 1991; Normark et al., 1993; Normark et al., 1998) have added further emphasis to the study of turbidite systems highlighting the relationships between eustasy-driven sea level changes and turbidite deposition as well as those between these systems and the other component systems tracts of depositional sequences.

Tectonism and its control on the growth of turbidite systems has been amply discussed by several authors (e.g., de Vries Klein, 1985; Stow et al. 1985; Allen and Homewood, 1986; Mutti et al., 1988; Pickering et al., 1989).

More recently, Milliman and Syvitski (1992), Mulder and Syvitski (1995) and Mutti et al. (1996) have highlighted the importance of flood-generated hyperpycnal flows associated with high-gradient fluvial systems of tectonically active basins and their role in generating turbidity currents. This clearly implies that the analysis of turbidite systems has to be framed within the broader stratigraphic context of the entire basin fill (see later).

7 — GENERAL CHARACTERISTICS OF TURBIDITE SYSTEMS OF THRUST-AND-FOLD BELTS

7.1 — Turbidite systems and their main component elements

The term "turbidite system" was introduced at the COMFAN of 1982 (Bouma et al., 1985) to avoid confusion between modern depositional systems characterized by a fan-shaped plan-view geometry and ancient depositional systems whose plan-view geometry is generally difficult to determine. Despite the many attempts in this direction, defining a turbidite system remains difficult and largely depends upon the set of data available and the approach taken.

Mutti (1985) and Mutti and Normark (1987, 1991) have suggested that the fill of turbidite basins can be sub-divided into hierarchically-ordered units which, from the largest to the smallest, include turbidite complex, system, stage and sub-stage (Fig. 16). Taking this approach, an ancient turbidite system is considered as a mappable stratigraphic unit composed of genetically linked facies associations recording a major forestepping-backstepping episode of basinal turbidite sand deposition. Although systems may differ considerably from each other, these authors also suggested that turbidite systems can be essentially viewed as composed of three basic and intergradational types which are shown in Fig. 17.

Distinctive facies associations and erosional surfaces define mappable elements that record the main transfer and depositional zones of submarine sediment gravity flows within each system. Large-scale submarine erosional features and channels are the characteristic transfer zones of turbidite systems; laterally extensive sandstone lobes and associated basin-plain deposits represent the depositional zones of these systems. Along basin margins, thick wedges of slope mudstones with thin-bedded and fine-grained turbidites form the link between basinal turbidite deposition and fluvio-deltaic systems formed in shelfal and nearshore regions (see later). These elements and their vertical and lateral stratigraphic relationships are summarized in the diagrams of Fig. 18a and 18b.

7.2 — General depositional setting

Peripheral basins of thrust-and-fold belts are large and elongate features stretched parallel to the local structural axes and are commonly filled with point-sourced turbidite systems. Classic examples of such settings are those of the Miocene Marnoso-arenacea in the northern Apennines (Ricci Lucchi, 1986) and the Eocene Hecho Group in the south-central Pyrenees (Mutti, 1985).

The general depositional setting of these basins is relatively simple, particularly in the south-central Pyrenees where excellent exposures and good stratigraphic control make it possible to trace these turbidite basin-fills into their slope, shelfal, and fluvio-deltaic equivalents (e.g., Mutti et al., 1985, 1988, 1996), probably a very unique setting in this respect.

Basically, basin-fills of this kind show large-scale erosional features deeply cut into slope and shelfal strata of basin margins that can be traced basinward into thick and extensive accumulations of sandstone lobes and terminal basin-plain deposits. The lateral extent and volume of the sediment contained in basin-fills of this kind vary greatly from one basin to another, mainly depending on basin size and configuration and sediment availability. Most commonly, basin-fills of this type have lengths in excess of 100 km and thicknesses up to as much as 4000 m. Flexural subsidence related to thrust-loading and uplift of hinterland regions clearly play a major role here in creating space for sedimentation and sediment to fill in this space respectively.

The general interpretation suggested by settings of this type is shown in Fig. 9. Submarine gravity flows, originated from basin margins, first accelerated on steep erosional conduits and then traveled considerable distances across the basin floor depositing turbidite sandstone beds of remarkable individual volume and lateral continuity.

How can we relate settings of this type with the processes and facies we discussed in previous sections? Clearly, this is only possible if we have good data about facies distribution patterns and the stratigraphy of the basin fill considered. To this purpose we will avail ourselves once more of information derived from the Eocene Hecho Group of the south-central Pyrenees and consider the main facies types and sandstone-body geometry that can be observed from the fill of large-scale erosional features down to the most distal basin-plain regions. These facies have been described in detail in a number of papers (Mutti, 1977, 1992; Mutti and Normark, 1987, 1991; Mutti et al., 1988; Remacha et al., 1998) and are further examined in these notes. For the reader's convenience, a general stratigraphic cross section of the Eocene fill of the south-central Pyrenean foreland basin is shown in Fig. 19.

7.3 — Large-scale submarine erosional features

Examples of large-scale erosional features from the south-central Pyrenees are shown in Figs. 20 and 21. Detailed stratigraphic analysis and mapping show that the width of these features was in the order of several km and their relief could reach as much as 500 m, thus making them comparable to the canyons and fan valleys of many modern fan systems (see Normark et al., 1993, for an updated review of these features in both modern and ancient turbidite systems). Their relief decreases in a basinward direction, where these features are replaced by shallow channels which are, in most cases, difficult to recognize in the field.

Although the main and genetically independent fill of these features is predominantly associated with subsequent progradation of fine-grained slope wedges, relatively thin units of very coarse grained deposits commonly occur in their axial portions. These deposits are of great importance because they record the first phase of deposition of gravity flows along their conduits. Figure 22 shows part of the axial fill of the Gerbe-Cotefablo system in the Eocene south-central Pyrenees, one of the best examples of this kind of sedimentation. These deposits form an overall fining-upward succession that grades from basal clast-supported, boulder- to coarse pebble-sized conglomerates (F3 deposits of Fig. 8), forming poorly-defined downstream accreting bars and lenses, into progressively finer-grained and vertically-accreting facies which are capped by mudstones (Mutti et al., 1985). The basal conglomerates are interpreted as the coarsest deposit left behind by bypassing granular flows and reworked by the same and/or subsequent flows.

It is difficult to assess the importance of these deposits without taking into account two other significant aspects of the problem. The first is the amount of submarine erosion which is associated with these deposits, implying that very large volumes of fine-grained sediment were necessarily incorporated into gravity flows that were moving downslope (Fig. 9); the second is that huge volumes of basinal sandstone lobes and basin-plain deposits occur farther downcurrent of these conglomerates (see cross section of Fig. 19).

A setting of this kind strongly suggests that submarine conduits of gravity flows experienced substantial erosion by these flows and contributed large volumes of fines to them. The scheme in Fig. 9 summarizes this process assuming that (1) granular flows are accelerated along their conduits because of steep slopes, and (2) large-scale bed erosion takes place in the head of these flows. Bed erosion is generated by turbulence and the impact of the coarsest particles, which collect at the head of the flow (see above), on a muddy substratum. Mud is eroded as rip-up clasts which are partly incorporated within the basal granular flow and most commonly thrown backward where they will concentrate and float along the boundary between the basal flow and the overlying turbulent flow (see experimental results of Postma et al., 1988 for large floating clasts in bipartite turbidite flows). Disaggregation of these water-impregnated and soft clasts during the downslope motion of the flow results in mud particles which become incorporated as suspended load in the upper turbulent flow, thus increasing its density and velocity. The upper turbulent flow thus progressively thickens and accelerates down the conduit since the positive acceleration resulting from bed erosion (increased density) must exceed the negative acceleration imparted by the mixing with ambient water. This phase of bed erosion and acceleration of the flow is probably that conceptually postulated by Parker (1982) for the ignition of catastrophically erosive turbidity currents and their transition to a self-sustained state. During this phase, the granular flow travels ahead of the turbulent flow.

Spectacular evidence for this kind of process is provided by Fig. 23 showing the example of a carbonate megaturbidite of the Hecho Group basin in the south-central Pyrenees. At least eight carbonate megaturbidites, whose origin is likely to be associated with seismic activity (seismoturbidites of Mutti et al., 1984), are interbedded with the basinal terrigenous turbidites of the Hecho Group. Some of these units have individual thicknesses in excess of 200 m, runout distances up to 100 km and dry sediment volume in excess of 200 km^3 (Johns et al., 1981; Labaume et al., 1987). Because of their exceptional volume and the relatively limited extent of the receiving basin, these flows could not spread enough across their final depositional zone in order to produce a significant horizontal segregation of their sediment load. As a result, the final deposit of each flow is an overall graded unit whose depositional divisions are shown in Fig. 23. The basal megabreccia division is made up of huge shelfal carbonate blocks that outdistanced finer carbonate debris during the catastrophic motion of a granular flow. Large-scale submarine bed erosion of the head of the flow is documented by extra-large calcareous mudstone clasts forming a very distinctive division above the megabreccia and representing an original slope deposit. The mudstone-clast division is overlain by a normally graded carbonate turbidite which is capped by a thick division of calcareous mudstone. The latter is thought to result from the disaggregation of rip-up calcareous mudstone clasts. The megabreccia divisions contain impressive features related to the catastrophic dewatering of an overpressured flow.

The example discussed above strongly suggests that gravity flows can actually originate as granular flows and that the origin of turbulent flows largely depends upon the amount of bed erosion that granular flows can produce at their front when moving on relatively steep slopes mainly driven by inertia forces.

7.4 — Sandstone lobes and basin-plain deposits

As discussed above, large-scale submarine erosional features are an extremely important erosive element of turbidite systems since, along their length, gravity flows accelerate and become high-momentum, fast-moving flows. Clearly, from a sedimentation standpoint, these submarine conduits are zones which are mainly characterized by erosion and sediment bypass. The depositional zones of these flows are located farther downcurrent in lobe and basin-plain regions, where gradual deceleration forces these flows to deposit their sand and mud load.

At the exit of their conduit, inertia flows enter progressively flatter and less confined basinal regions essentially as jet flows. Enhanced mixing with ambient fluid due to lateral spreading and friction causes the progressive deceleration of the flow which is thus bypassed by the upper turbulent flow and sheared by it from above. During this process, the basal flow progressively loses its overpressured fluid, and its finer-grained sediment becomes incorporated into the overlying turbulent flow. Elutriated inertia flows are thus forced to deposit their coarser-grained sediment load due to increased shearing resistance and the more distal deposits of these flows are typically reworked and winnowed by bypassing turbulent flows. Characteristic tractional features, commonly expressed by 3D megaripples, may form in these distal and still relatively coarse-grained deposits. Further downcurrent, turbulent flows progressively decelerate in space and time, depositing the bulk of grain populations C and D, mainly through traction-plus-fallout and fallout processes.

Sandstone lobes and associated basin-plain deposits form a continuum recording different stages during the main phase of deposition of turbidity currents. These sediments are characterized by an impressive tabular geometry over tens and hundreds of kilometers along basin axes. For this reason, they have also been referred to as "sheet systems" (Pickering et al., 1989).

In thrust-and-fold belt basins, lobe and basin-plain deposits are represented by huge volumes of alternating sandstones and mudstones sometimes forming the bulk of an entire orogenic belt, as in the case of the northern Apennines. Originally defined by Mutti and Ricci Lucchi (1972), these turbidite elements are sometimes difficult to distinguish, particulary in oversupplied basins. In general, however, sandstone lobes have a considerably higher sand/shale ratio and record the region where the great majority of turbidity currents deposit their coarser sand load. Basin-plain strata are generally finer-grained and may be commonly interbedded with hemipelagic layers.

Mutti and Johns (1978) suggested that basin-plain turbidites would be deposited by very-large volume flows that could bypass the entire lobe region, thus representing highly catastrophic and infrequent flows. More recently, Remacha et al. (1998) have shown for the Hecho Group of the southern Pyrenees that nearly 65% of the flows that deposit in the lobe region can reach the basin plain. The remaining 35% of the turbidity currents deposit thin beds in the lobe region but cannot reach the basin plain because of their relatively small volume and momentum.

An example of basinal sandstone lobes from the Eocene Banaston system of the south-central Pyrenees is shown in the cross section of Fig. 24. The cross section shows a stratigraphic unit with a maximum thickness of some 1000 m in its proximal sector that can be traced basinward for more than 80 km. The setting of these sediments needs further comment. Paleocurrents are directed toward the viewer near the "high" on the east (the Boltana anticline, see Mutti et al., 1988 for more details); farther westward, paleocurrent directions are essentially toward WNW. The cross section clearly shows the impressive lateral continuity of tabular lobe packages over a distance of about 60 km. However, it is clear from the cross section that these bodies are considerably thicker near the "high" than away from it. This thickness change is here interpreted as the result of flow deceleration against the "high" and therefore of the control of submarine structurally-induced topography on sedimentation. This is an excellent example of what Kneller (1995) illustrated in his laboratory experiments.

Another example of turbidite sandstone lobes is that illustrated in the cross section of Fig. 25, showing the stratigraphically upper part of the Eocene Broto system in the south-central Pyrenees. The cross section shows the remarkable lateral continuity of sandstone beds over a distance of about 60 km in a direction strictly parallel to paleocurrent direction (from right to left). In this case, basin topography was very smooth, thus permitting the gradual deceleration of turbidity currents with distance along an essentially flat basin-floor surface. In such a downcurrent setting, individual sandstone beds progressively thin out, sand/shale ratio decreases, and broad channeling features smooth out and disappear. Detailed facies analysis of these strata indicates that F5 beds deposited by the overpressured basal granular flow progressively pass into F6 and F7 beds, indicating conditions of substantial traction, and eventually into F8 beds recording sedimentation primarily controlled by high rates of fallout from waning and depletive flows when starting their final stage of deceleration (see above).

These beds are very impressive in terms of variations of internal sedimentary structures and grain-size, strongly supporting the conclusion by Mutti (1992, p.105) that "no two turbidite beds are really the same". These variations are largely due to bypass features that occur almost within every bed, generating vertical breaks in grain size and sharp vertical transitions between tractional features, "en masse" deposition, and classic traction-plus-fallout laminae. These bypass features indicate that even in their main depositional zone most turbidity currents still contain grain populations that can be transported farther downstream.

The examples of the cross sections of Figs. 26 and 27 show the transition between lobe and basin-plain deposits and the characteristics of typical basin-plain strata respectively. As clearly indicated by Fig. 26, distal lobes and basin plain deposits form a continuum over a distance of approximately 20 km and it is difficult to separate these two types of turbidite facies associations without very careful detailed bed-by-bed correlations. It will be noted that the sediments shown in this cross-section are correlative with the thick and laterally continuous sandstone lobes of Fig. 25. Basically, basin-plain turbidites are characterized by thicker mudstone divisions produced by ponding in the distal part of the basin and by the very common occurence of internal depositional structures generated by turbidity currents repeatedly reflected and deflected from basin-margin slopes. These features, first described by Pickering and Hiscott (1985) from the Cloridorme Formation of Quebec, give way to very complex vertical successions of depositional divisions within turbidite beds. These features and their origin have been discussed in detail by Remacha et al. (1998). The cross section of Fig. 27 shows an impressive example of bed-by-bed correlation of basin-plain strata over a distance of some 27 km. Note the relatively thick mudstone divisions and the occurrence of thin hemipelagic intercalations.

7.5 — *Some remarks on the terms "channels" and "lobes"*

In their most simple and unambiguous expression, the terms "channel" and "lobe" refer to an individual flow and its deposit. For each flow considered, there is a zone where the flow originates, a transfer zone along which the flow and its sediment load move downslope, and finally a depositional zone where the flow undergoes deceleration and deposits the bulk of its load. The extent, as well as both the plan-view and the cross-sectional geometry, of the final deposit of an individual flow depends on many local controlling factors.

Consider the ideal example of Fig. 28 depicting a very highly-efficient flow. It will be noted that the transfer zone of the flow is first confined and erosive, and sedimentation is restricted to residual gravel (F3). Then the flow begins to spread laterally and deposition of F6 sediment occurs before the final deposit (F7 through F9) forms farther downcurrent. Basically, we have included F3 and F6 deposits in the transfer zone of the flow relative to the final depositional zone. What do the terms "channel" and "lobe" mean in this case? Essentially, the terms mean the transfer and the depositional zone of each flow respectively. Almost everybody would agree that "lobe" means deposition and therefore a bed characterized by some lateral extent. The concept of "channel" is more difficult to define. Most sedimentologists would relate this term to erosion and therefore to a convex-downward geometry of a deposit; fewer would think in terms of sediment bypass or transfer zone. This is a major communication problem for sedimentologists. For this reason, in previous sections we have emphasized transfer and depositional zones — concepts which are apparently clearer for everybody.

The pattern of Fig. 28 is exactly the plan-view expression of the facies tract shown in Fig. 7b and 12b. All other things being equal, this pattern will be maintained, with minor modifications, as long as the volume of the flows remains roughly constant. If subsequent flows are of larger volume, the pattern of Fig. 28 will change: the transfer zone will become longer and a larger lobe will form farther downcurrent. The opposite will occur if subsequent flows are of smaller volume. For a large-volume flow, the lobe deposit will form in basinal distal regions and the deposit will be far away from the origin of the flow. For a small-volume flow, the lobe deposit will form near the origin of the flow and usually be contained within large-scale conduits or channels acting as transfer zones for large-volume flows. It follows from the above that the concepts of channel and lobe should be treated with great caution. Broad and shallow channels, virtually undetectable in the field, may act as transfer zones of large-volume flows in basinal regions; on the other hand, lobe sediments deposited by small-volume flows may form within channels.

The relationships between the volume of individual turbidity currents and the location of their depositional zones have been well documented by

Normark et al. (1979) for the modern Navy fan, off California. The basinward or landward shifting of the transfer and depositional zones of turbidity currents through time as a function of their volume is an extremely important controlling factor of the cyclic stacking patterns displayed by turbidite systems (Mutti, 1985; Mutti and Normark, 1987; Mutti et al., 1994, with references therein). This problem is briefly reconsidered in the final part of these notes.

From a geological standpoint, channels and lobes cannot be obviously defined in the simple and somewhat conceptual way we have discussed above. Channels and lobes have to refer to mappable elements of a turbidite system and therefore to long-lived features which account for the transfer and deposition of large volumes of sediment through time. The reader is referred to Pickering et al. (1989), Mutti and Normark (1991), Normark et al. (1993), Clark and Pickering (1996) and Richards et al. (1998) for more details and different opinions, particularly as far as channels are concerned. We emphasize here that most ancient sandstone-filled submarine channels actually represent the lobes of relatively small-volume flows and that these fills cannot be used to predict substantial sand accumulations farther basinward. Channels, if strictly intended as transfer zones, can only be defined on their facies types indicating substantial sediment bypass and regardless of the geometry. We also emphasize that most outcrop-scale identified channels are in fact scours produced by gravity flows. Mistaking a scour produced by an individual flow for a transfer zone of a large number of turbidity currents may have some unpleasant consequences in terms of oil exploration and production.

Many of these problems may be partially overcome if we restrict the terms "transfer zone" and "depositional zone" to a genetic facies approach and use more descriptive terms like "channel", "channel fill", "channel deposit", "lobe" and "sheet sandstone" to denote the geometry of a sandstone body without genetic implications.

8 — SEDIMENTARY CYCLICITY AND SEQUENCE STRATIGRAPHY OF TURBIDITE SYSTEMS

8.1 — General

Turbidite systems of orogenic belts are characterized by a spectacular sedimentary cyclicity that developed on physically and temporally different scales, ranging from that of basin fills to that of m-thick packages of sandstone beds. In particular, the best preserved and thus the most complete expression of these cyclic stacking patterns is observed in turbidite sandstone lobe successions which are characterized by hierarchically-ordered alternations of thick-bedded and predominantly sandstone facies with muddier units (the classic facies sequences of Mutti and Ricci Lucchi, 1972, 1975; see Pickering et al., 1989, for an extensive discussion).

This kind of cyclic turbidite sedimentation has long been considered as primarily related to autocyclic processes (e.g., fan progradation or recession, channel-avulsion and lobe-switching, etc.) inherent to deep sea fan depositional dynamics (e.g., Mutti and Ricci Lucchi, 1972; Walker, 1978). More recently, attempts have been made to explain this kind of sedimentation also in terms of cyclic relative sea-level variation (e.g., Mutti and Normark, 1987; Pickering et al., 1989; Vail et al., 1991) or of tectonically-induced instability processes in basin-margin regions (e.g., Mutti, 1992). However, mounting evidence from different fields of research, directly or indirectly suggests that cyclic turbidite sedimentation may have a considerably more complex origin involving other potentially important controlling factors such as tectonism, climate, sediment flux to the sea, and the fluvial regime in adjacent mountain fronts (e.g., Milliman and Syvitski, 1992; Mutti et al., 1994, 1996; Mulder and Syvitski, 1995). A better understanding of turbidite cyclic stacking patterns may thus contribute significant information regarding not only the evolution of orogenic belt basins but also the triggering mechanisms of submarine gravity flows and their timing.

Fig. 29 shows examples of hierarchically-ordered cyclic stacking patterns from the Eocene Broto turbidites in the south-central Pyrenees. The cyclicity developed in a sandstone lobe succession, that is in strata recording the main depositional zone of turbidite systems and thus very probably characterized by the highest preservation potential of the different depositional events through which these systems grew up through time (see Schumm, 1977, for the same concept developed for the depositional zones, alluvial or marine, of fluvial systems).

As shown by Mutti et al. (1996), an almost identical type of cyclicity is observed in the depositional zones of ancient flood-dominated fluvio-deltaic systems of orogenic belts basins. In these zones, sedimentation is recorded by cyclically-stacked shelfal sandstone lobes deposited by catastrophic hyperpycnal flows (see later).

Whatever the scale, in both turbidite and flood-dominated fluvio-deltaic systems the cyclicity is expressed by forestepping-backstepping sandstone facies and records a similarly cyclic basinward and landward shifting of the sandy depositional zones in basinal, shelfal, shallow-marine and alluvial environments. This poses two basic problems which, unfortunately, are still very poorly understood: 1) the nature of the causative processes of this kind of sedimentation, and 2) the reasons why these processes operate cyclically through time.

In the following pages, we briefly and tentatively discuss these problems. However, before venturing further into this discussion, it seems useful to briefly review the main concepts related to flood-dominated sedimentation and their implications for an understanding of turbidite sediments in orogenic belts.

8.2 — Floods and turbidity currents

Some fundamental papers have recently dealt with the importance of hyperpycnal plumes generated at the mouths of many modern rivers (e.g., Milliman, and Syvitsky, 1992; Mulder and Syvitsky, 1995). These flows are increasingly considered as an important process in marine delta construction, particularly in tectonically active settings characterized by small- and medium-sized mountainous rivers with high-elevation catchment basins and therefore high sediment yield. In these rivers, termed "dirty rivers" by Mulder and Syvitski (1995), sediment flux to the sea is high because of the elevation of drainage basins and their proximity to the shoreline. These factors strengthen the role of flooding since sediment concentration of flood-generated flows is not reduced by energy dissipation and sedimentation in extensive alluvial flood basins and coastal plains, thus favouring the formation of hyperpycnal flows in adjacent sea waters. In most modern settings, however, these hyperpycnal flows can only carry fine-grained sediment to the sea.

These relatively small rivers have been shown to be of fundamental importance in the origin of ancient fan-delta and river-delta systems of many tectonically active basins (Mutti et al., 1996). Detailed stratigraphic reconstructions and facies analyses indicate that flood-generated hyperpycnal flows actually built up huge accumulations of conglomerate, pebbly-sandstone, sandstone, and mudstone facies extending across marginal-marine and shelfal/slope regions. The volume of sediment involved in this kind of catastrophic sedimentation is difficult to perceive on the basis of what one can observe in and predict from modern environments.

The marine depositional zones of ancient flood-dominated fluvio-deltaic systems are typically represented by cyclically-stacked shelfal sandstone lobes, i.e. packages of sharp-based and graded sandstone beds commonly containing hummocky cross stratification produced under combined-flow conditions inherent to flood-related processes. The stratigraphic and sedimentological importance of these sandstone lobes as related to flood-dominated fluvio-deltaic systems, first perceived by Martinsen (1990) for the Namurian of northern England, has been documented by Mutti et al. (1996). Sediments of this kind form huge accumulations of nearshore and shelfal regions in many basins worldwide, spanning in age from Paleozoic to Pleistocene. These sediments probably represent the most genuine expression of fluvial-dominated delta-front deposits since, in the absence of flood-generated hyperpycnal flows, river-borne sands can only be redistributed in marine environments by waves and tides and related currents.

Since ancient flood-dominated fluvio-deltaic systems are commonly associated with important volumes of deeper water turbidite sandstones, Mutti et al. (1996) have suggested that there must be a genetic relationship between the two kinds of sedimentation and that in tectonically active settings a common trigger of turbidity currents must be related to flood-generated hyperpycnal flows entering sea water.

The way in which hyperpycnal flows can be directly or indirectly related to the origin of turbidity currents in modern basins has been discussed in several papers (e.g. Normark and Piper, 1991; Zeng et al., 1991; Mulder et al., 1998). There appears to be the common tendency to consider that hyperpycnal flows entering sea waters may generate sediment overloading and failure at river mouths, thus triggering slumps and turbidity currents. However, Mutti et al. (1996) have shown that in many ancient flood-dominated fluvio-deltaic systems hyperpycnal flows may directly transform into turbidity currents through a phase of extensive bed erosion and bulking in shallow marine and shelfal environments. These currents may travel considerable distances as sustained flows if sediment is continuously supplied from upcurrent through long-lived hyperpycnal flows. Depending on the local physiographic setting, and particularly on the width and slope of the shelf, these flows will deposit their sediment load in shelfal regions (flood-generated shelfal lobes) or directly in adjacent deep-water basins (turbidites).

8.3 — Large-scale cyclic stacking patterns

These patterns are observed in basin-fill successions and are developed on physical and temporal scales roughly coinciding with those of 3rd-order depositional sequences of seismic and sequence stratigraphy. An example of this cyclicity is shown in the cross-section of Fig. 19 depicting part of the infill of the Eocene south-central Pyrenean foreland basin through turbidite and flood-dominated fluvio-deltaic systems. Available biostratigraphic data (Mutti et al., 1988; Waehry, 1999; Serra-Kiel et al., 1994; Barbera et al., 1997) indicate that this cyclicity roughly matches that of the 3rd-order cycles of sequence stratigraphy.

Sequence-stratigraphic models restrict turbidite sedimentation to periods of lowstand of sea-level during which former shelfal regions undergo subaerial exposure and fluvial erosion (Mitchum, 1985; Vail, 1987; Posamentier et al., 1988). In these models, which have been devised particularly for divergent margin settings, turbidites would form when rivers debouch directly onto the slope, thus favouring the formation of gravity flows. A basal, sand-rich "basin-floor fan", commonly characterized by a pronounced mounded geometry, records this first phase. This fan is subsequently capped by a thick slope wedge, or "slope fan", related to the progradation of lowstand fluvio-deltaic complexes. The slope fan is generally mud-rich and essentially coincides with the well known channel-levee complexes of many modern fans (e.g., Weimer, 1990).

A great number of papers have dealt with sequence stratigraphy of turbidite systems with strong emphasis on their seismic expression in passive-margin basins (see a synopsis in Weimer and Link, 1991, with references therein). Most of this work clearly shows that many turbidite systems are associated with low sea-level conditions, particularly in the case of Pliocene and Pleistocene systems whose growth has been strongly controlled by high-frequency eustatic oscillations (e.g. Weimer, 1990; Normark et al., 1998).

Although the term "relative sea-level variations" is clearly intended to mean a combination of eustasy and tectonism (Posamentier et al., 1988), in reality, sequence-stratigraphic models heavily rely on eustatic cycles developed at different hierarchical orders and assume that the classic depositional sequences are the product of a 3rd-order, eustasy-driven cyclicity. The interested reader is referred to the recent paper by Dewey and Pitman (1998) for an extensive discussion of this problem. The main conclusion of these authors is that eustasy-driven cyclicity obviously exists but not in the range of the 3rd-order cycles advocated by sequence-stratigraphic models. Consequently, the authors suggest that third order cycles could be essentially related to tectonism (see also Sloss, 1988). A very similar conclusion was reached by Mutti et al. (1996) through the study of the sedimentary cyclicity of turbidite systems of orogenic belts. The boundaries of turbidite-bearing depositional sequences are invariably associated with pronounced angular unconformities associated with phases of thrust propagation and tectonic uplift. In these settings, both turbidite and associated flood-dominated fluvio-deltaic systems primarily record periods of time during which sediment flux to the sea is dramatically increased by tectonic uplift along basin margins in order to increase subaerial erosion and ensuing sediment availability.

The scheme of Fig. 30 offers a sequence-stratigraphic interpretation of turbidite-bearing depositional sequences of tectonically active basins which differs substantially from that of eustasy-driven models of passive margins. The fundamental difference resides in the fact that in tectonically active basins the boundaries of depositional sequences are produced by phases of structural deformation and associated uplift of basin-margin regions. As already mentioned, the origin of such sequence boundaries can be clearly documented for the Eocene south-central Pyrenean foreland basin (Mutti et al., 1985, 1988). The scheme of Fig. 30 suggests that tectonic uplift, denudation and subsidence control the gross geometry of depositional sequences, i.e. their accomodation space and sediment volume. The scheme also suggests that climate is the main trigger of floods and related hyperpycnal flows, whose magnitude would decrease with time as a result of decreasing elevation of fluvial drainage basins and related sediment availability (see later).

Briefly, the proposed model is based on the degree of catastrophism of floods (Fig. 31). Highly catastrophic floods are generated by the steep gradients of fluvial systems and large sediment availability in alluvial basins, as well as along basin margins. Under these conditions, floods enter seawater as powerful and highly erosive hyperpycnal flows which transform in sustained turbidity currents that can reach deep and basinal regions. This direct link between fluvial and turbidite sedimentation gives way to a kind of depositional system which has been referred to as "fluvio-turbidite system" (Fig. 32; Mutti et al., 1996). In our interpretation, most thrust-and-fold belt turbidite accumulations record this style of catastrophic sedimentation.

8.4 — Small-scale cyclic stacking patterns

There seems to exist a general consensus on the fact that, within third order depositional sequences, shallow-marine and coastal strata are characteristically arranged in high-frequency cyclic stacking patterns recording hierarchically-ordered cycles in accommodation. These stacking patterns, typically expressed on the scale of parasequences, are generally interpreted as the product of 4th- and 5th-order, eustasy-driven sea-level variations (e.g.; Posamentier et al. 1988; Mitchum and Van Wagoner, 1991). How these sequence-stratigraphic concepts can be applied to fluvial sedimentation poses a series of problems concerning the readjustment of fluvial systems and changes of sediment flux to the sea in response to baselevel variations (e.g., Schumm, 1993; Wescott, 1993). Similar problems are encountered in interpreting the cyclicity of turbidite successions in terms of mere, eustasy-driven accommodation cycles.

Sequence-stratigraphic models heavily rely upon "normal" marine processes, such as waves and tides, which are obviously related to cyclic variations in water depth. To what extent these variations may actually affect sedimentation dominated by hyperpycnal flows and turbidity currents remains difficult to understand, particularly in outer shelfal, slope and basinal regions where subsidence tends to continuously create space for sedimentation.

Parasequence-scale or high-frequency cyclicity is one of the most conspicuous features of turbidites particularly in their sandy depositional zones (see above). Examples of this high-frequency cyclicity are shown on the right part of Fig. 31. Basically, this cyclicity records forestepping-backstepping episodes of sand deposition which are repeated in hierarchically-ordered cycles. Each cycle is expressed by a relatively abrupt basinward shift of the sandy depositional zones of the turbidity currents followed by a relatively gradual backstepping of these zones until sand deposition ceases and muddier facies are deposited. This style of cyclic deposition, associated with the shift of transfer and depositional zones through time, is discussed in more detail elsewhere (Mutti et al., 1994). A very similar high-frequency cyclicity characterizes the depositional zones of flood-dominated fluvio-deltaic system, thus suggesting a clear genetic link between these two types of sedimentation and a common causative process (Mutti et al., 1996).

The diagrams of Figs. 33a and 33b compare the basic forestepping-backstepping facies tracts that can be observed in turbidite and flood-dominated fluvio-deltaic systems. The physical and temporal scales involved are those of the parasequences. These diagrams clearly show a first phase of forestepping of both transfer and depositional zones of the flows resulting from the sudden increase in the flow volume. Most commonly, in proximal zones this forestepping stage is recorded by sharp and locally erosive surfaces. More distally, this sharp surface may be replaced by a transitional contact showing a distinct thickening- and coarsening-upward trend of a limited number of beds. The backstepping stage is characteristically gradual and records the progressive landward migration of the transfer and depositional zones of individual flows due to their similarly progressive decrease in volume.

At this point, it seems difficult to escape the preliminary conclusion that tectonism and climate are the basic controlling factors of both turbidite and fluvio-deltaic deposition in orogenic belts. The importance of tectonism has been discussed above. Climate and its cyclic variations become fundamental for an interpretation of small-scale cyclic stacking patterns generated by floods and related hyperpycnal flows and turbidity currents. An excellent and well-documented example of climatic control on turbidite deposition has been described by Weltje and de Boer (1993) for cyclically stacked sandstone lobes in the Pliocene of Corfu, Greece.

High-frequency (Milankovitch range) cyclic climatic variations are needed to make available large amounts of water in fluvial drainage basins over short periods of time in order to produce floods. Heavy rainfall, snow and ice melting, and breeching of naturally-dammed lakes are the processes commonly envisaged to produce catastrophic floods (e.g., Baker and Bunker, 1985; Costa, 1988; Costa and Schuster, 1988). Once they have been generated, floods move has water-sediment mixtures that may evolve in different ways depending upon their sediment concentration and the local physiographic setting (Mutti et al., 1996).

High-frequency cyclicity of both fluvio-deltaic and turbidite sediments (Figs. 33a and 33b) clearly shows an alternation of relatively coarse- and fine-grained facies suggesting periods of floods alternating with periods of "normal" sedimentation, i.e. a climatic cyclicity. During periods characterized by floods and related processes (hyperpycnal flows and sustained turbidity currents), there is a very distinct trend, within each facies sequence, suggesting that the volume of individual flows declined after an initial peak. Such a trend, which is almost ubiquitous in both turbidite and flood-dominated fluvio-deltaic systems, could be explained through the progressive decrease of the amount of water made available in fluvial drainage basins through time or, more likely, from the combined effect of the above with sediment availability. In other words, much of the sediment available is incorporated in the first floods, thus lowering the concentration and the efficiency of subsequent flows.

Geomorphic cycles (uplift, erosion and subsidence) punctuated by high-frequency climatic fluctuations (Milankovitch cyclicity) seem to govern much of the terrigenous sedimentation of orogenic belt basins. The most complete record of these different types of cyclicity is probably contained in the deep-water turbidite systems of these basins and is still waiting to be unravelled. The very simple conclusion of these notes is that problems of turbidite sedimentation require, first of all, a broad approach to the tectonic history of the basin under consideration.

AKNOWLEDGEMENTS

E. Mutti aknowledges financial support provided by the Italian Ministry of Public Education and the Consiglio Nazionale delle Ricerche, Italy.

REFERENCES

ALLEN P.A. & ALLEN J.R. (1990) — *Basin analysis principles and applications.* Black. Sci. Publ., 454 pp., Oxford.

ALLEN P.A., & HOMEWOOD P. N. (Eds.) (1986) – *Foreland Basins.* I.A.S. Spec. Publ. N° 8, Blackwell Scientific, 453 pp., Oxford.

BAKER V.R. & BUNKER R.C. (1985) – *Cataclysmic late Pleistocene flooding from glacial lake Missoula: a review.* Quater. Sci. Rev., v. 4, pp. 1-41, Oxford.

BARBERÀ X., MARZO M., REGUANT S., SAMSÒ J.M., SERRA-KIEL J. & TOSQUELLA J. (1997) — *Estratigrafìa del Grupo Fìgols (Paleògeno, cuenca de Graus-Tremp, NE de Espaublic*Rev. Soc. Geol. de Espa Fìgols (Paleògeno, cuenca de Graus-

BOUMA A.H. (1962) — *Sedimentology of some flysch deposits, a graphic approach to facies interpretation.* Elsevier Co. 168 pp., Amsterdam.

BOUMA A.H., NORMARK W.R. & BARNES N.E. (Eds.) (1985) — *Submarine fans and related turbidite systems.* Springer-Verlag, 351 pp, Berlin Heidelberg.

CHAN M.A. & DOTT R.H.Jr. (1983) — *Shelf and deep-sea sedimentation in Eocene forearc basin, Western Oregon — Fan or No Fan?*

A.A.P.G. Bull., v. 67, pp. 2100-2116, Tulsa.

CLARK J. D. & PICKERING K. T. (1996) – *Submarine channels, processes and architecture.* Vallis Press Eds., 231 pp.

COSTA J.E. (1988) – *Reologic, geomorphic and sedimentologic differentiation of water floods, hyperconcentrated flows and debris flows.* In: BAKER V.R., KOCHEL R.C. & PAITON P.C. (Eds.), Flood Geomorphology. Wile-Interscience Publ., pp. 113-122, New York.

COSTA J.E. & SCHUSTER R.L. (1988) – *The formation and failure of natural dams.* Geol. Soc. Am. Bull., v. 100, pp. 1054-1068, Boulder.

DE VRIES KLEIN G. (1985) – *The frequency and periodicity of preserved turbidites in submarine fans as a quantitative record of tectonic uplift in collision zones.* In: CARTER N.L. & UYEDA C. (Eds.), Collision Tectonics: deformation of continental lithosphere.
Tectonophysics, v. 119, pp. 181-193, Oxford.

Dewey J.F. & Pitman W.C. (1998) — *Sea-level changes: mechanisms, magnitudes and rates.* In: Pindell J.L. & Drake S. (Eds.), Paleogeographic evolution and non-glacial eustacy.
S.E.P.M. Spec. Publ. N°58, pp. 1-16, Tulsa.

Dzulynski S., Ksiazkiewicz M. & Kuenen Ph.H. (1959) — *Turbidites in flysch of the Polish Carpathian Mountains.*
Geol. Soc. Am. Bull., v. 70, pp. 1089-1118, Boulder.

Faugères J.C. & Stow D.A.V. (1993) – *Bottom-current-contolled sedimentation: a synthesis of the contourite problem.*
Sedimentary Geology, v. 82, pp. 287-297, Amsterdam.

Fisher R.V. (1983) — *Flow trasformations in sediment gravity flows.*
Geology, v. 11, pp. 273-274, Boulder.

Friedman G.M. & Sanders J.E. (1997) — *Dispelling the myth of sea-floor tranquillity.* Geotimes, v. january 1997, pp. 24-27, Tulsa.

Gee M.J.R., Masson D.G., Watts A.B. & Allen P.A. (1999) – *The Saharan debris flow: an insight into the mechanics of long runout submarine debris flows.* Sedimentology, v. 46 (2), pp. 317-335, Oxford.

Gladstone C., Phillips J.C. & Sparks R.S.J. (1999) – *Experiments on bidisperse, constant-volume gravity currents: propagation and sediment deposition.* Sedimentology, v. 45 (5), pp. 833-843, Oxford.

Harms J.C. & Fahnestock R.K. (1965) — *Stratification, bedform and flow fenomena (with an exemple from the Rio Grande).* In: Middleton G.V. (Ed.), Primary Structures and Their Hydrodinamyc Interpretation.
S.E.P.M. Spec. Publ. n°12, pp. 84-115, Tulsa.

Heller P.L. & Dickinson W.R. (1985) — *Submarine ramp facies model for delta-fed, sand-rich turbidite systems.* A.A.P.G. Bull., v. 69 pp. 960-976, Tulsa.

Inverson R.M. (1997) – *The physics of Debris Flows.*
Rev. Geoph., v. 35, pp. 245-296.

Johns D. R., Mutti E., Rosell J., Seguret M. (1981) – *Origin of a thick, redeposited carbonate bed in Eocene turbidite of the Hecho Group, south-central Pyrenees, Spain.* Geology, v. 9, pp. 161-168, Boulder.

Kneller B. (1995) — *Beyond the turbidite paradigm: physical models for deposition and their implications for reservoir prediction.* In: Hartlet A.J. & Prosser D.J. (Eds.), Characterization of Deep Marine Clastic Systems. Geol. Soc. Spec. Publ. N° 94, pp. 31-49, Boulder.

KNELLER B.C. & BRANNEY M.J. (1995) — *Sustained high-density turbidity currents and the deposition of thick massive sands.*
Sedimentology, v. 42, pp. 607-616, Oxford.

KUENEN Ph.H. (1950) — *Turbidity currents of high density.* Intern. Geol. Congr., 18th, London, 1948, Rept. 8, pp. 44-52.

KUENEN Ph.H. (1957) — *Sole markings of graded graywacke beds.*
Journ. Geol., v. 65, pp. 231-258, Chicago.

KUENEN Ph.H. (1964) — *Deep sea sands and ancient turbidites* In: BOUMA A.H. & BROUWER A. (Eds.), Turbidites. Elsevier Co., pp. 3-33, Amsterdam.

KUENEN Ph.H. & MIGLIORINI C.I. (1950) — *Turbidity currents as a cause of graded bedding.* Journ. Geol., v. 58, pp. 91-127, Chicago.

KUENEN Ph.H. & HUMBERT F.L. (1964) — *Bibliography of turbidity currents and turbidites.* In: BOUMA A.H. & BROUWER A. (Eds.), Turbidites. Elsevier Co., pp. 222-246, Amsterdam.

LABAUME P., MUTTI E. & SEGURET M. (1987) — *Megaturbidites: a depositional model from the Eocene of the SW-Pyrenean foreland basin, Spain.*
Geo-Marine Letters, v. 7, pp. 91-101, Berlin Heidelberg.

LOWE D.R. (1982) — *Sediment gravity flows: II. Depositional models with special reference to the deposits of high-density turbidity currents.*
Journ. Sed. Petr., v. 52 (1), pp. 279-297, Tulsa.

LUTHI S. (1981) — *Experiments on non-channelized turbidity currents and their deposits.* Mar. Geol., v. 20, pp. M59-M68, Amsterdam.

MARR J., HARFF P., SHANMUGAM G. & PARKER G. (1997) – *Experiments on subaqueous sandy debris flows.* Supplement to EOS Transactions, AGU Fall Meeting, San Francisco, v. 78 (46), F347.

MARTINSEN O. J. (1990) — *Fluvial, inertia-dominated deltaic deposition in the Namurian (Carboniferous) of northern England.*
Sedimentology, v. 37 (6), pp. 1099-1114, Oxford.

MIDDLETON G.V. (1966a) — *Experiments on density and turbidity currents.I. Motion of the head.* Canadian Jour. Earth Sci.,v. 3, pp. 523-546, Ottawa.

MIDDLETON G.V. (1966b) — *Experiments on density and turbidity currents.II. Uniform flow of density currents.*
Canadian Jour. Earth Sci., v. 3, pp. 627-637, Ottawa.

MIDDLETON G.V. (1967) — *Experiments on density and turbidity currents.III. Deposition of sediment.*
Canadian Jour. Earth Sci.,v. 4, pp. 475-505, Ottawa.

MIDDLETON G.V. & HAMPTON M.A. (1973) — *Sediment gravity flow: mechanics of flow and deposition.* In: MIDDLETON G.V. & BOUMA A.H. (Eds.),Turbidites and deep-water sedimentation; Pacific section, S.E.P.M. Short Course Notes n°1, pp. 1-38, Tulsa.

MIGLIORINI C.I. (1943) — *Sul modo di formazione dei complessi tipo Macigno.*
Boll. Soc. Geol. It., v. 62, pp. 48-50, Roma.

MILLIMAN J.D. & SYVITSKI J.P.M. (1992) — *Geomorphic and tectonic control of sediment discharges to the ocean: the importance of small mountain rivers.* Jurn. Geol., v. 100 (5), pp. 525-544, Chicago.

MITCHUM R.M.Jr. (1985) — *Seismic stratigraphic expression of submarine fans.* In: BERG O.R. & WOOLVERTON D.G. (Eds.), Seismic Stratigraphy II: An Integrated Approach To Hydrocarbon Exploration.
A.A.P.G. Mem. N° 39, pp. 117-136, Tulsa.

MITCHUM R. M. & VAN WAGONER J. C. (1991) – *High-frequency sequences and their stacking patterns: sequence-stratigraphic evidence of high-frequency eustatic cycles.*
Sedimentary Geology, V. 70, pp. 131-160, Amsterdam.

MOHRIG D., WHIPPLE K.X., HONDZO M., ELLIS C. & PARKER G. (1998) – *Hydroplaning of subaqueous debris flows.*
Geol. Soc. Am. Bull., v. 110, pp. 387-394, Boulder.

MULDER T. & SYVITSKI J.P.M. (1995) — *Turbidity currents generated at river mouths during exceptional discharges to the world oceans.*
Journ. Geol., v. 103 (2), pp. 285-299, Chicago.

MULDER T. & SYVITSKI J.P.M. (1996) — *Climatic and morphologic relationships of rivers: implications of sea-level fluctuactions on river loads.* Journ. Geol., v. 104 (5), pp. 509-542, Chicago.

MULDER T., SYVITSKI J.P.M. & SKENE K.I. (1998) — *Modeling of erosion and deposition of turbidity currents generated at river mouths.*
Journ. Sed. Res., v. 68, pp. 124-137, Tulsa.

MUTTI E. (1977) — *Distinctive thin bedded turbidite facies and related depositional environments in the Eocene Hecho Group (South-Central Pyrenees, Spain).* Sedimentology, v. 24, pp. 107-131, Oxford.

Mutti E. (1979) — *Turbidites et cones sous-marine profonds.* In: Homewood P. (Ed.), Sedimentation Detritique (Fluviatile, Littorale et Marine), Institut de Geologie, Université de Fribourg, pp. 353-419.

Mutti E. (1985) — *Turbidite systems and their relations to depositional sequences.* In: Zuffa G.G. (Eds.), Provenance of Arenites: NATO-ASI Series, Reidel Publishing Co., pp. 65-93.

Mutti E. (1992) — *Turbidite sandstones.* AGIP — Istituto di Geologia, Università di Parma, 275 pp., San Donato Milanese.

Mutti E., Davoli G., Mora S. & Papani L. (1994) — *Internal stacking patterns of ancient turbidite systems from collisional basins.* In: WEIMER P., Bouma A.H. & Perkins B. (Eds.), Submarine Fans and Turbidite Systems, Papers Presented at the GCS S.E.P.M. 15th Annual Research Conference, pp. 257-268, Austin.

Mutti E., Davoli G, Tinterri R. & Zavala C. (1996) — *The importance of fluvio-deltaic systems dominated by catastrophic flooding in tectonically active basins.* Mem. Sci. Geol., v. 48, pp. 233-291, Padova.

Mutti E. & Ghibaudo G. (1972) — *Un esempio di torbiditi di conoide esterna: le Arenarie di S. Salvatore (Formazione di Bobbio, Miocene) nell' Appennino di Piacenza.*
Mem. Acc. Sci. Torino, Cl. Sci. Fis. Mat. Nat. N° 16, pp. 1-40, Torino.

Mutti E. & Johns D.R. (1978) — *The role of sedimentary by-passing in the genesis of fan fringe and basin plain turbidites in the Hecho Group System (South-Central Pyrenees).* **Mem. Soc. Geol. It., v. 18, pp. 15-22, Roma.**

Mutti E. & Normark W.R. (1987) — *Comparing examples of modern and ancient turbidite systems: Problems and Concepts.* In: Legget J.K. & Zuffa G.G. (Eds.), Marine Clastic Sedimentology: Concept and case studies, Graham & Trotman, pp. 1-38, London.

Mutti E. & Normark W. R. (1991) — *An integrated approach to the study of turbidite systems.* In: Weimer P. & Link H. (Eds.), Seismic Facies and Sedimentary Processes of Submarine Fans and Turbidite Systems, pp. 75-106, Ann Arbor.

Mutti E., Remacha E., Sgavetti M., Rosell J., Valloni R, & Zamorano M. (1985) – *Stratigraphy and facies characteristic of the Eocene Hecho Group turbidite systems, south-central Pyrenees.* In: Mila M. D. & Rosell J. (Eds.), Excursion Guidebook: VI Eur. Reg. Mtg. I.A.S., Lerida, Spain, Excursion 12, pp. 521-576.

Mutti E. & Ricci Lucchi F. (1972) — *Le torbiditi dell' Appennino settentrionale: introduzione all'analisi di facies.*
Mem. Soc. Geol. It., v. 11, pp. 161-199, Roma.

Mutti E. & Ricci Lucchi F. (1974) — *La signification de certaines unité séquentielles dans les séries à turbidites.*
Bull. Soc. Geol. Fr., série 7, XVI (6), Paris.

Mutti E. & Ricci Lucchi F. (1975) — *Turbidite facies and facies association.* In: Mutti E., Parea G.C., Ricci Lucchi F., Sagri M., Zanzucchi G., Ghibaudo G. & Iaccarino S. (Eds.), Examples of Turbidite Facies Associations from Selected Formation of Northern Apennines, IX Int. Cong. I.A.S., Field Trip, All., pp. 21-36, Nice, France.

Mutti E. Ricci Lucchi F., Seguret M. & Zanzucchi G. (1984) — *Seismoturbidites: a new group of risedimented deposits.* In: Cita M. B. & Ricci Lucchi F. (Eds.), Seismicity and sedimentation, Elsevier Scientific Publ., pp. 103-116, Amsterdam.

Mutti E., Seguret M. & Sgavetti M. (1988) — *Sedimentation and deformation in the Tertiary sequences of the southern Pyrenees.* A.A.P.G. Mediterranean Basin Conference, Spec. Publ. Institute of Geology, University of Parma, Field trip N° 7, 153 pp., Parma.

Nardin T.R., Hein F.J., Gorsline D.S. & Edwards B.D. (1979) — *A review of mass movement processes, sediment and acoustic characteristics and contrasts in slope and base-of-slope systems versus canyon-fan-basin floor systems.* In: L.J. Doyle & O.H. Jr Pilkey (Eds.), Geology of Continental Slopes. S.E.P.M Spec. Publ. N° 27, pp. 61-73, Tulsa.

Norem H., Locat J. & Schieldrop B. (1990) — *An approach to the physics and the modeling of submarine flowslides.*
Marine Geotechnology, v. 9, pp. 93-111, London.

Normark W.R. (1970) — *Growth patterns of deep-sea fans.*
A.A.P.G. Bull., v. 54, pp. 2170-2195, Tulsa.

Normark W.R., Mutti E. & Bouma A.H. (Eds.) (1983/1984) — *Submarine clastic systems: Deep sea fans and related turbidite facies.*
Geo-Marine Letters, v. 3, pp. 53-224, Berlin Heidelberg.

Normark W.R. & Piper D.J. (1991) — *Initiation processes and flow evolution of turbidity currents: Implications for the depositional record.* S.E.P.M. Spec. Publ., N° 46, pp.207-230, Tulsa.

NORMARK W.R., PIPER D.J.W. & HESS G.R. (1979) — *Distributary channels, sand lobes, and mesotopography of Navy submarine fan, California borderland, with applications to ancient fan sediments.* Sedimentology, v. 26, pp. 749-774, Oxford.

NORMARK W.R., PIPER D.J.W. & HISCOTT R.N. (1998) — *Sea level controls on the textural characteristics and depositional architecture of the Hueneme and associated submarine fan systems, Santa Monica Basin, California.* Sedimentology, v. 45, pp. 53-70, Oxford.

NORMARK W. R., POSAMENTIER H. & MUTTI E. (1993) — *Turbidite systems: state of the art and future directions.* Rev. Geoph., v. 31, pp. 91-116.

PAREA G.C. (1965) — *Evoluzione della parte settentrionale della geosinclinale appenninica dall' Albiano all' Eocene superiore.*
Mem. Acc. Sc. Lett. Art. Modena, v. 7, pp. 1-98, Modena.

PARKER G. (1982) — *Condition for the ignition of catastrophically erosive turbidite currents.* Mar. Geol., v. 46, pp. 307-327, Amsterdam.

PICKERING K.T. & HISCOTT R.N. (1985) – *Contained (reflected) turbidity currents from the middle Ordovician Cloridorme formation, Quebec, Canada: an alternative to the antidune hypothesis.*
Sedimentology, v. 32, pp. 373-394, Oxford.

PICKERING K.T., HISCOTT R.N. & HEIN F. J. (EDS.) (1989) — *Deep marine environments: clastic sedimentation and tectonics.*
Unwin Hyman Ltd, pp. 416.

PICKERING K. T., STOW D.A.V., WATSON M. P. & HISCOTT R.N. (1986) — *Deepwater facies, processes and models: a review and classification scheme for modern and ancient sediments.*
Earth Science Reviews, v. 23, pp. 75-174.

POSAMENTIER H.W., ERSKINE R.D. & MITCHUM R.M Jr. (1991) — *Submarine fan deposition within a sequence stratigraphic framework.* In: WEIMER P. & LINK H. (Eds.), Seismic facies and sedimentary processes of submarine fans and turbidite systems.
Springer-Verlag, pp. 75-106, Berlin Heidelberg.

POSAMENTIER H.W., JERVEY M.T. & VAIL P.R. (1988) — *Eustatic controls of clastic deposition I — Conceptual framework.* In: WILGUS C.K., HASTINGS B.S., KENDALL C.G.ST.C., POSAMENTIER H.W., ROSS C.A. & VAN WAGONER J.C. (Eds.), Sea level changes: an integrated approach.
S.E.P.M. Spec. Publ. N° 42, pp. 109-124, Tulsa.

POSTMA G., NEMEC W. & KLEINSPEHN K.L. (1988) — *Large floating clasts in turbidites: a mechanism for their emplacement.*
Sedimentary Geology, v. 58, pp. 47-61, Amsterdam.

PRIOR D.B. & BORNHOLD B.D. (1989) — *Submarine sedimentation on a developing Holocene fan delta.*
Sedimentology, v. 36, pp. 125-143, Oxford.

RAVENNE C. & BEGHIN P. (1983) — *Apport des expériences en canal à l'interprétation sédimentologique des dépôts de cônes détritiques sous-marins.* Revue de l'I.F.P., v. 38 (3), pp. 279-297, Paris.

READING H. G. & RICHARDS M. T. (1994) – *The classification of deep-water siliciclastic depositional systems by grain size and feeder system.*
A.A.P.G. Bull., v. 78, pp. 792-822, Tulsa.

REMACHA E., FERNÀNDEZ L. P., MAESTRO E., OMS O. & ESTRADA R. (1998) — *The upper Hecho Group turbidites and their vertical evolution to deltas (Eocene South-Central Pyrenees) (complementary information to Field Trip a Guidebook).* 15th International Sedimentological Congress, April 8-12, 1998. Universidad Autònoma de Barcelona, Universidad de Oviedo. 54 pp.

RICCI LUCCHI F. (1986) — *The Oligocene to recent foreland basins of the northern Appennines.* In: ALLEN P.A. & HOMEWOOD P. (Eds.), Foreland Basins. I.A.S. Spec. Publ. n° 8, pp. 105-139, Oxford.

RICHARDS M., BOWMAN M. & READING H. (1998) — *Submarine-fan systems I: characterization and stratigraphic prediction.*
Mar. Petr. Geol., v. 15, pp. 689-717, Tulsa.

SANDERS J.E. (1965) — *Primary sedimentary structures formed by turbidity currents and related resedimentation mechanisms.* In: MIDDLETON G.v. (Ed.), Primary sedimentary structures and their hydrodinamic interpretation, S.E.P.M. Spec. Publ. N° 12, pp. 192-219, Tulsa.

SCHUMM S.A. (1977) – *The fluvial system.* JOHN WHILEY & SONS (Eds.), New York, pp. 338.

SCHUMM S.A. (1993) — *River response to baselevel change: implications for sequence stratigraphy.* Journ. Geol., v. 101, pp. 279-294, Chicago.

SERRA-KIEL J., CANUDO J.I., DINARES E., MOLINA N., ORTIZ J.O. PASCUAL J.M., SAMSO J.M. & TOSQUELLA J. (1994) — *Cronoestratigrafia de los sedimentos marinos del Terciario inferior de la Cuenca de Graus-Tremp (Zona Central Surpirenaica).*
Rev. Soc. Geol. de Espaos sedimentos marinos del Terciario i

SHANMUGAM G. (1996) — *High density turbidity currents: are they sandy debris flow?* Journ. Sed. Res., v. 66, pp. 2-10, Tulsa.

SHANMUGAM G. (1999) – *50 years of the turbidite paradigm (1950s-1990s): deep-water processes and facies models – a critical perspective.* Marine and Petroleum Geology (in press), Oxford6-08-1999 15:03 PM.

SHANMUGAM G., LEHTONEN L. R., STRAUME T., SYVERSTEN S. E., HODGKINSON R. J. & SKIBELI M. (1994) – *Slump and debris-flow dominated upper slope facies in the Cretaceous of the Norwegian and northern North Sea (61-67° N): implications for sand distribution.*
A.A.P.G. Bull., v. 78, pp. 910-937, Tulsa.

SHANMUGAM G., SPALDING T.D. & ROFHEART D.H. (1993) — *Process sedimentology and reservoir quality of deep-marine bottom-current reworked sands (sandy contourites): an example from the Gulf of Mexico.* A.A.P.G. Bull., v. 77, pp. 1241-1259, Tulsa.

SLOSS L.L. (1988) — *Forthy years of sequence stratigraphy.*
Geol. Soc. Am. Bull., v. 100 (11), pp. 1661-1665, Boulder.

STOW D.A.V., HOWELL D.G. & NELSON C.H. (1985) — *Sedimentary, Tectonic and Sea-Level controls.* In: BOUMA A.H., NORNARK W.R. & BARNES N. (Eds.), Submarine Fans and Related Turbidite Systems. Springer-Verlag, pp. 15-22, Berlin Heidelberg.

VAIL P.R. (1987) — *Seismic stratigraphy interpretation using sequence stratigraphy. Part 1: Seismic stratigraphy interpretation procedure.* In: BALLY A.W.(Eds.), Atlas of seismic stratigraphy. A.A.P.G. studies in Geology, v. 27, pp. 1-10, Tulsa.

VAIL P.R., AUDEMARD F., BOWMAN S.A., EINSER P.N. & PEREZ-CRUZ G. (1991) — *The stratigraphic signatures of tectonics, eustasy and sedimentation.* In: EINSELE G., RICKEN W. & SEILACKER A. (Eds.), Cycles and events in stratigraphy. Springer-Verlag, pp. 617-659, Berlin Heidelberg.

VIANA A.R., FAUGÈRES J.C. & STOW D.A.V. (1998) – *Bottom-current-contolled sand deposits – a review of modern shallow-to deep-water environments.* Sedimentary Geology, v. 115, pp. 53-80, Amsterdam.

WAEHRY A. (1999) – *Facies analysis and physical stratigraphy of the Ilerdian in the eastern Tremp-Graus Basin (south-central Pyrenees, Spain).* Terre & Environment, Inst. Forel, Dép. De Minéralogie, Dép. de Géologie et Paléontologie, Section des Sciences de la Terre, Université de Genève. v. 15, 191 pp.

WALKER R.G. (1967) — *Turbidite sedimentary structures and their relationship to proximal and distal depositional environments.*
Journ Sed. Petr., v. 37 (1), pp. 25-37, Tulsa.

WALKER R.G. (1978) — *Deep-Water Sandstones Facies and Ancient Submarine Fans: Models for Exploration for Stratigraphic Traps.*
A.A.P.G. Bull., v. 62, pp. 932-966, Tulsa.

WALKER R.G. & MUTTI E. (1973) — *Turbidite facies and facies associations.*
In: MIDDLETON G. v. & BOUMA A.H. (EDS.), Turbidites and deep water sedimentation, Pacific section.
S.E.P.M. Short Course notes, pp. 119-157, Tulsa.

WEIMER P. (1990) — *Sequence Stratigraphy, facies geometry and depositional history of the Mississippi Fan, Gulf of Mexico.*
A.A.P.G. Bull., v. 74, pp. 425-453, Tulsa.

WEIMER P. & LINK M.H. (Eds.) (1991) — *Seismic facies and Sedimentary Processes of Submarine Fans and Turbidites Systems.*
Springer-Verlag, 447 pp, Berlin Heidelberg.

WEIMER P., BOUMA A.H. & PERKINS B.F. (1994) — *Submarine Fans and Turbidite Systems, Sequence Stratigraphy, Reservoir Architecture and Production Characteristics Gulf of Mexico and International.* GCS S.E.P.M. Foundation 15th Annual Research Conference. 440 pp.

WELTJE G. & DE BOER P.L. (1993) – *Astronomically induced paleoclimatic oscillations reflected in Pliocene turbidite deposits on Corfu (Greece): Implications for the interpretation of higher order cyclicity in ancient turbidite systems.* Geology, v. 21, pp. 307-310, Boulder.

WESCOTT W.A. (1993) — *Geomorphic thresholds and complex response of fluvial systems — Some implications for sequence stratigraphy.*
A.A.P.G. Bull., v. 77, pp. 1208-1218, Tulsa.

ZENG J. & LOWE D.R. (1997a) — *Numerical simulation of turbidity current flow and sedimentation: I. Theory.*
Sedimentology, v. 44, pp. 67-84, Oxford.

ZENG J. & LOWE D.R. (1997b) — *Numerical simulation of turbidity current flow and sedimentation: II. Results and geological applications.*
Sedimentology, v. 44, pp. 85-104, Oxford.

ZENG J., LOWE D.R., WISEMAN W.J. & BORNHOLD B.D. (1991) — *Flow properties of turbidity currents in Bute Inlet, British Columbia.*
Sedimentology, v. 38, pp. 975-996, Oxford.

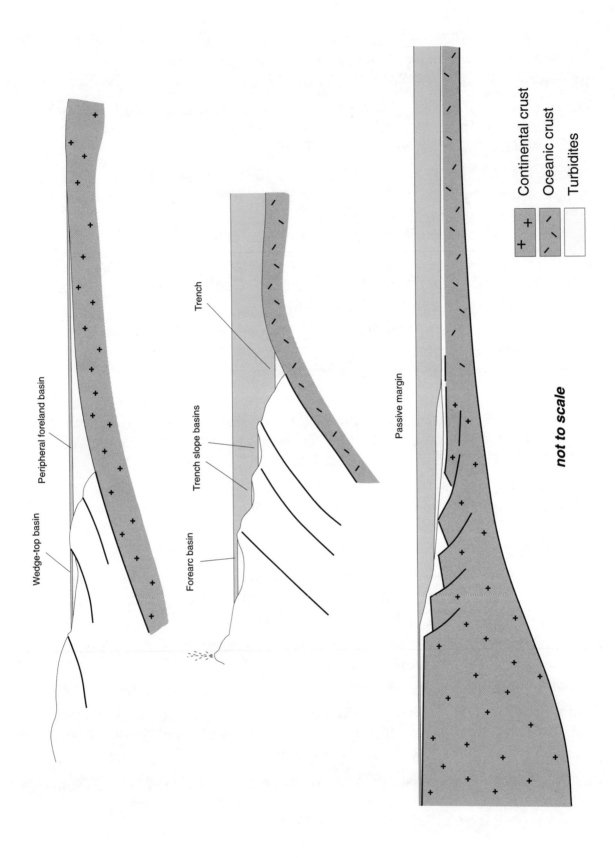

Fig. 1 - Turbidite basins in differents tectonic settings.

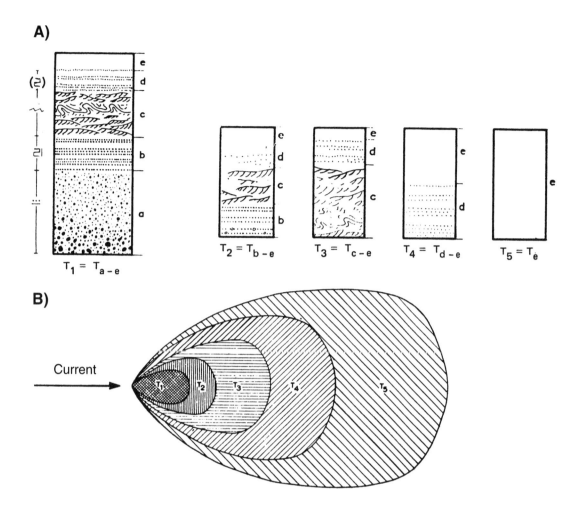

Fig. 2: The Bouma sequence (**A**) and its "depositional cone" (**B**) (from Bouma, 1962).

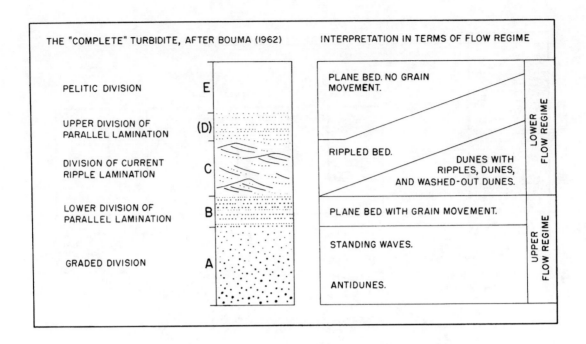

Fig. 3: The Bouma sequence and its interpretation (from Walker, 1967).

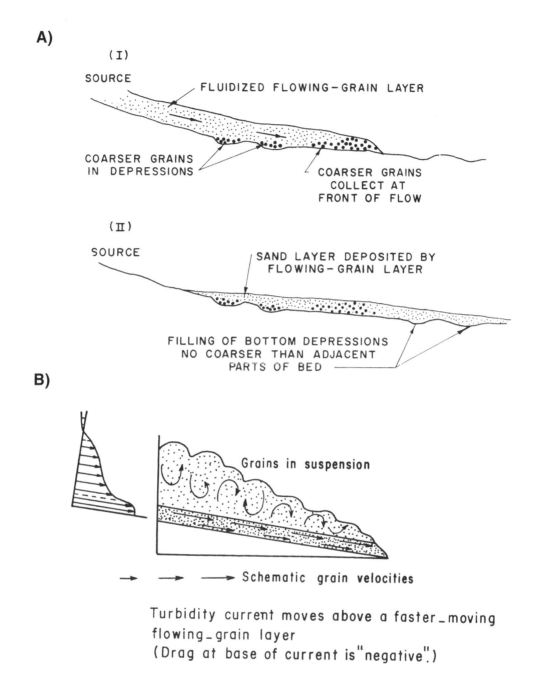

Fig. 4: A) Fluidized flowing grain layer.
B) Velocity profile of a turbidity current consisting of a basal faster moving flowing grain layer overlain by a turbulent flow (from Sanders, 1965).

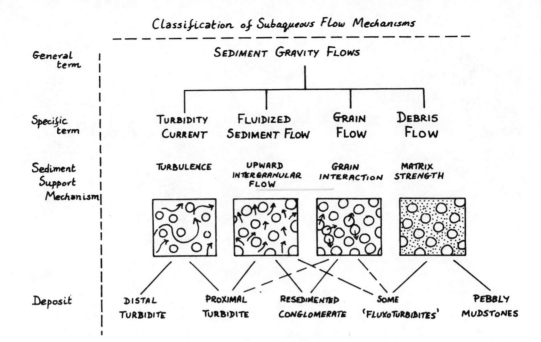

Fig. 5: Classification of subaqueous sediment gravity flows and particle support mechanisms (from Middleton and Hampton, 1973).

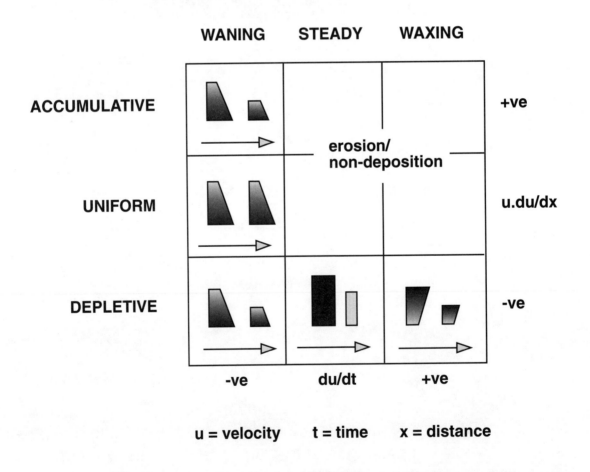

Fig. 6: Acceleration matrix, with illustrative bed sequences for each field showing downstream and vertical changes in relative grain size of the deposits of each field; arrows point downstream (from Kneller, 1995).

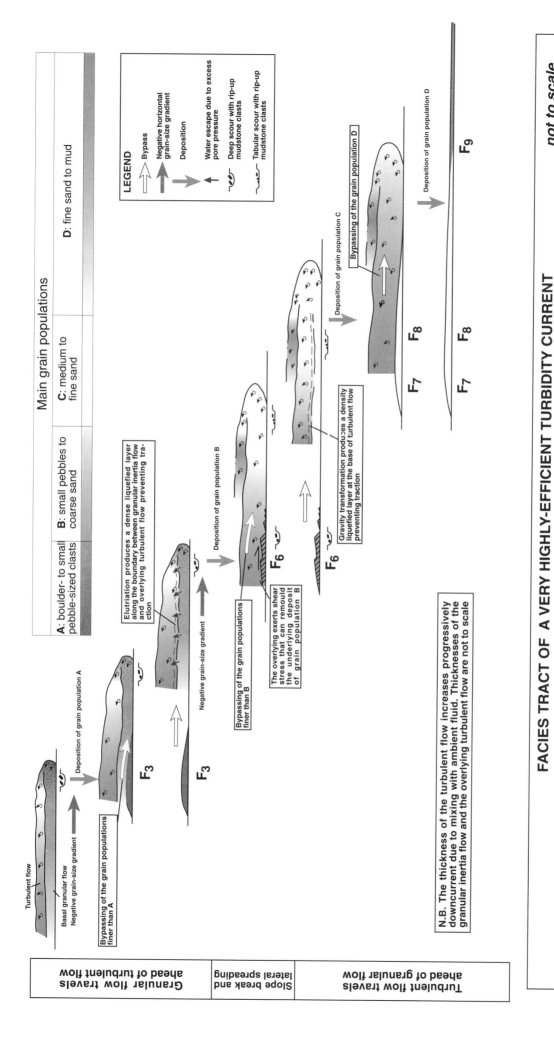

Fig. 7a: Diagram showing an ideal facies tract of a very highly-efficient turbidity current and main flow transformations occurring during downslope motion.

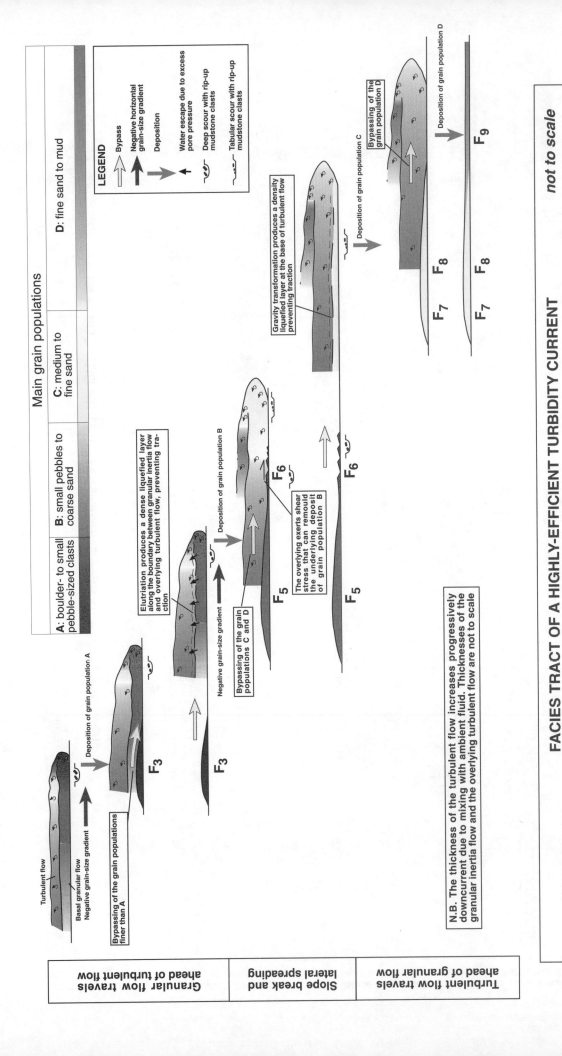

Fig. 7b: Diagram showing the facies tract of a highly-efficient turbidity current and main flow transformations occurring during downslope motion.

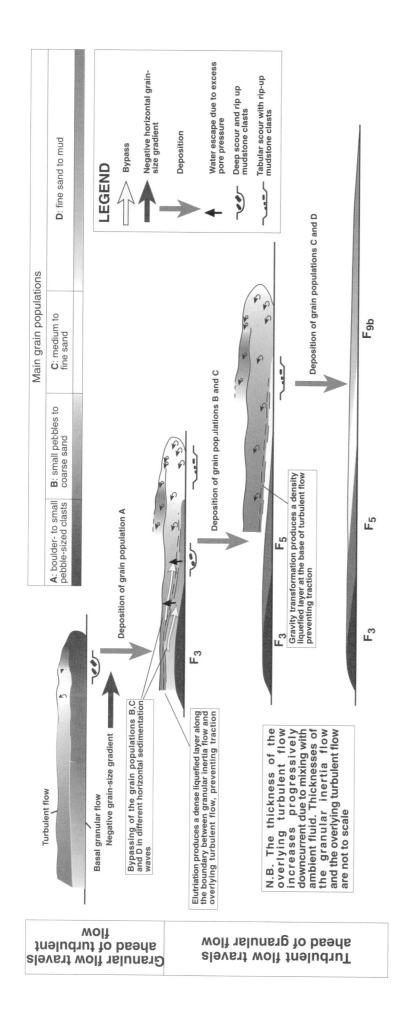

Fig. 7c: Diagram showing the facies tract of a very poorly-efficient turbidity current and main flow transformations occurring during downslope motion.

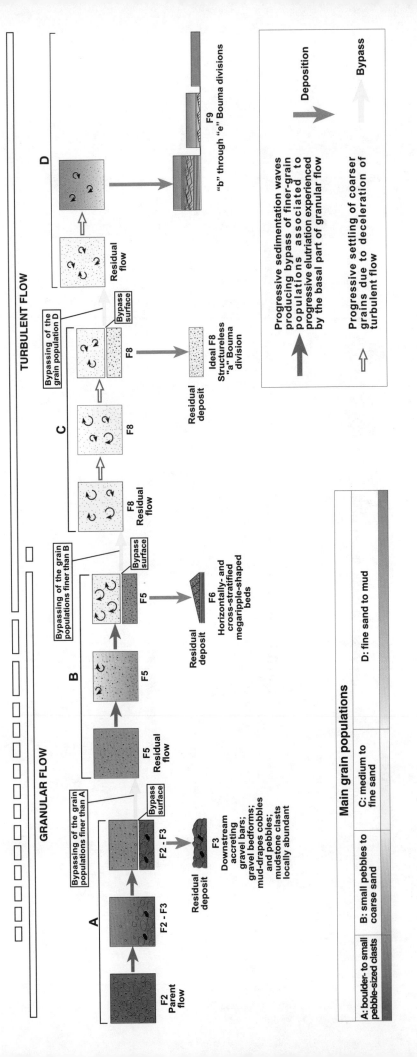

Fig. 8 - Facies and processes of highly-efficient turbidity currents.

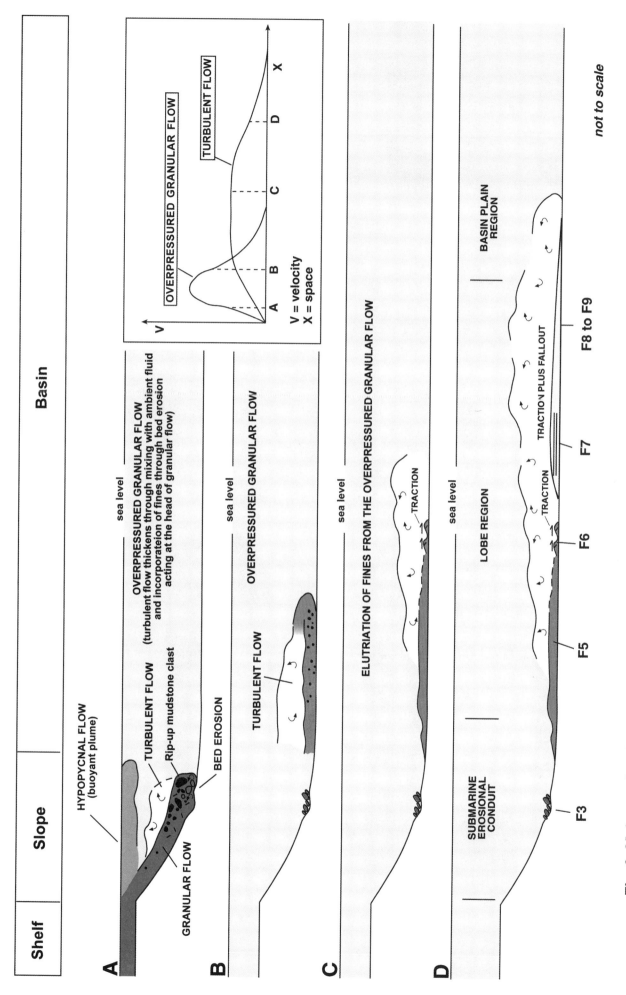

Fig. 9: Main erosional and depositional processes operating in a turbidite system dominated by highly efficient flows.

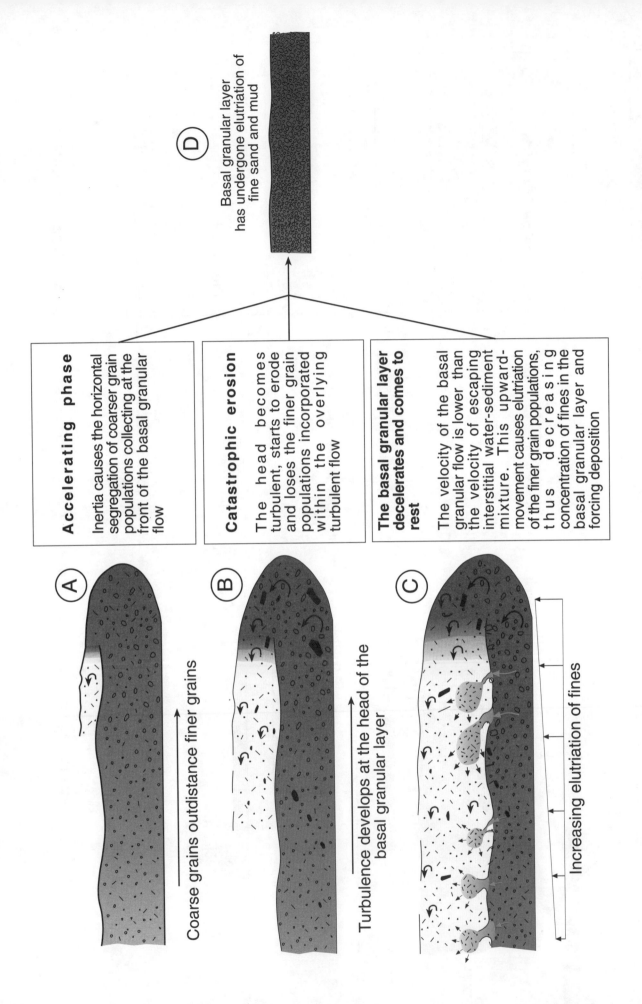

Fig. 10: Processes contributing to the elutriation of finer particles from the basal granular layer.

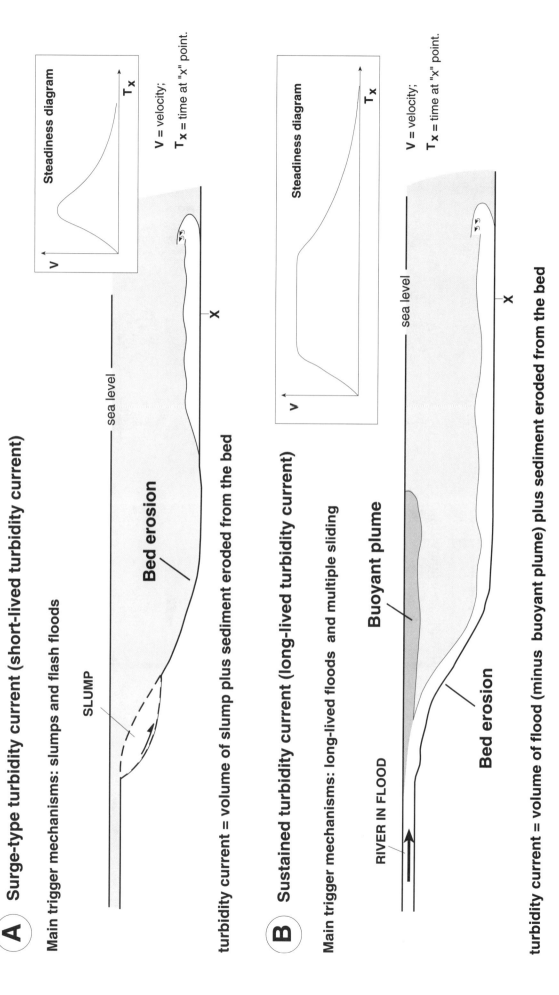

Fig. 11: Main types of turbidity currents. A spectrum of turbidity currents exists between these two end-members (**A** and **B**).

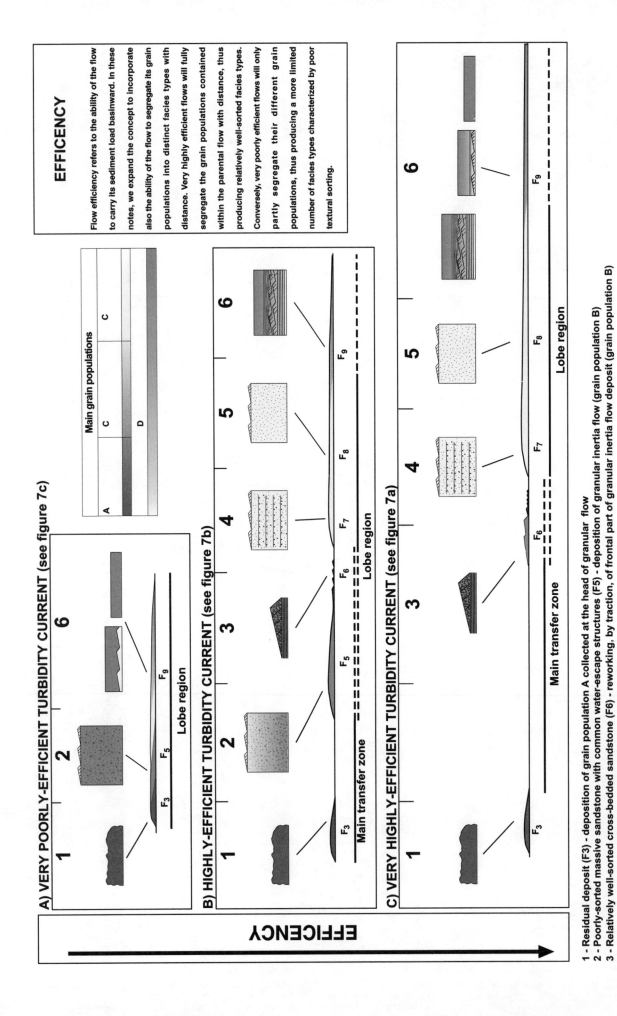

Fig. 12: Turbidite facies tracts as related to flow efficiency.

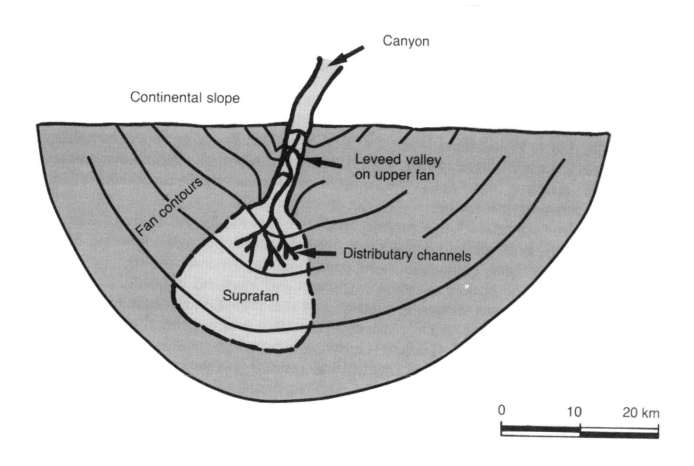

Fig. 13: Fan model of Normark (1970).

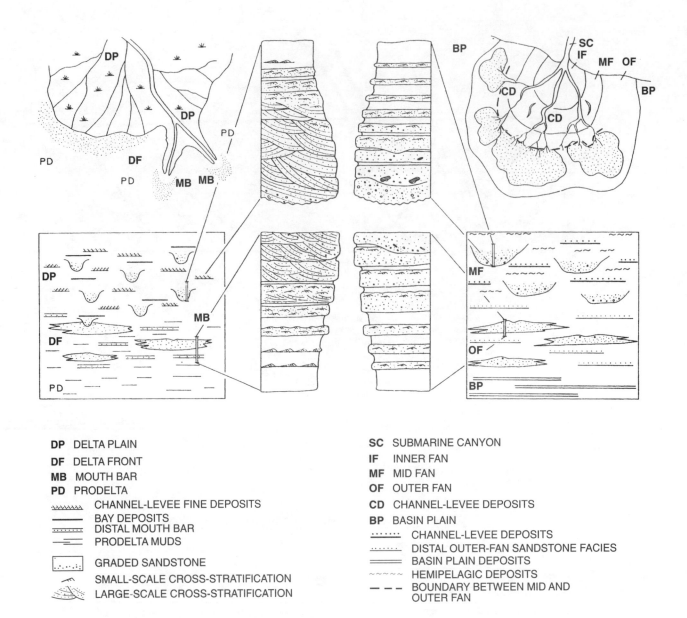

Fig. 14: Similarities between deltas and deep-sea fans (from Mutti and Ghibaudo, 1972).

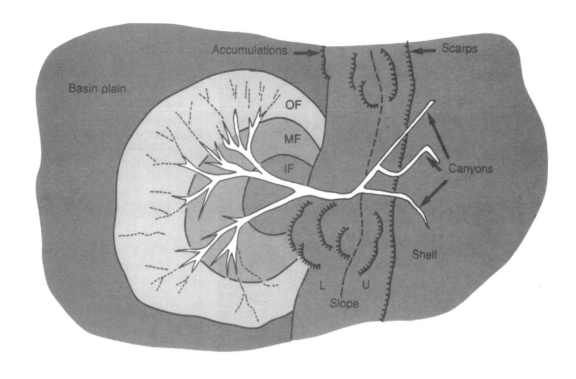

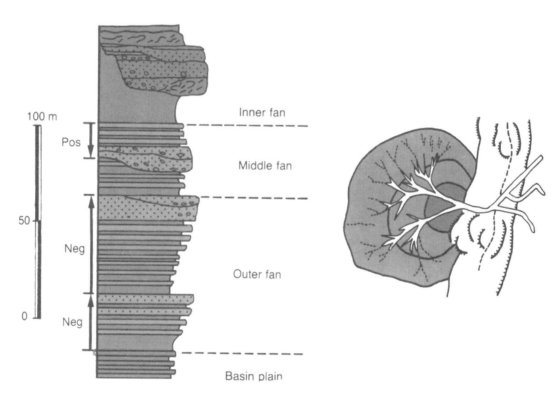

Fig. 15: Fan model of Mutti and Ricci Lucchi (1972).

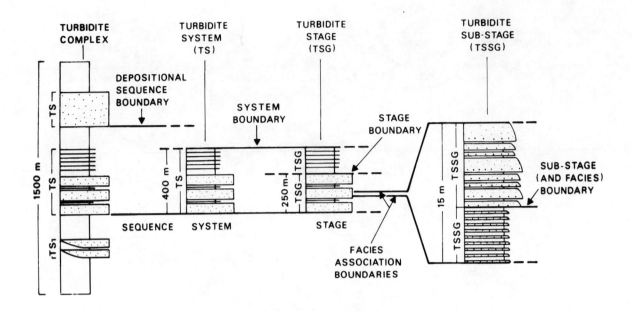

Fig. 16: Hierarchical classification of turbidite depositional units (from Mutti and Normark, 1987).

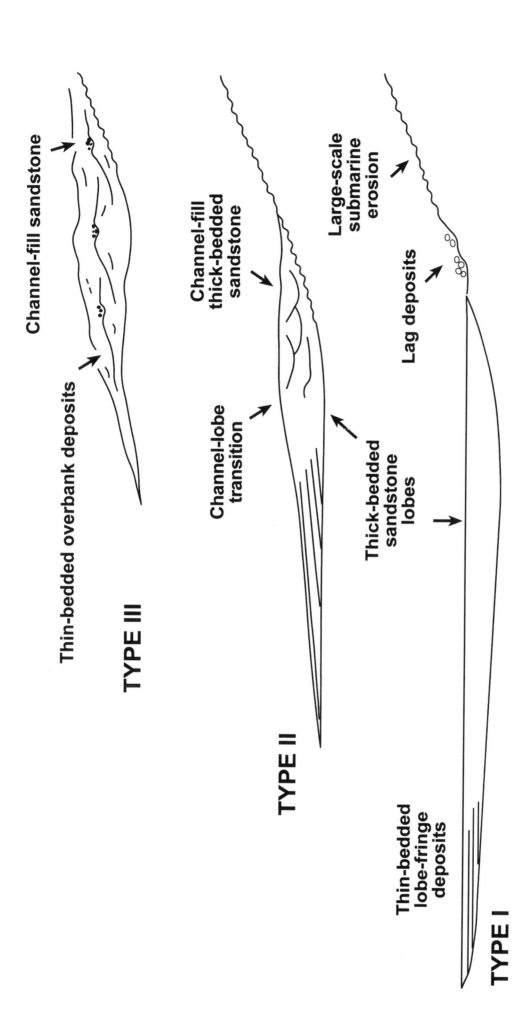

Fig. 17: Main types of turbidite depositional systems (from Mutti 1985).

BASIC TURBIDITE ELEMENTS

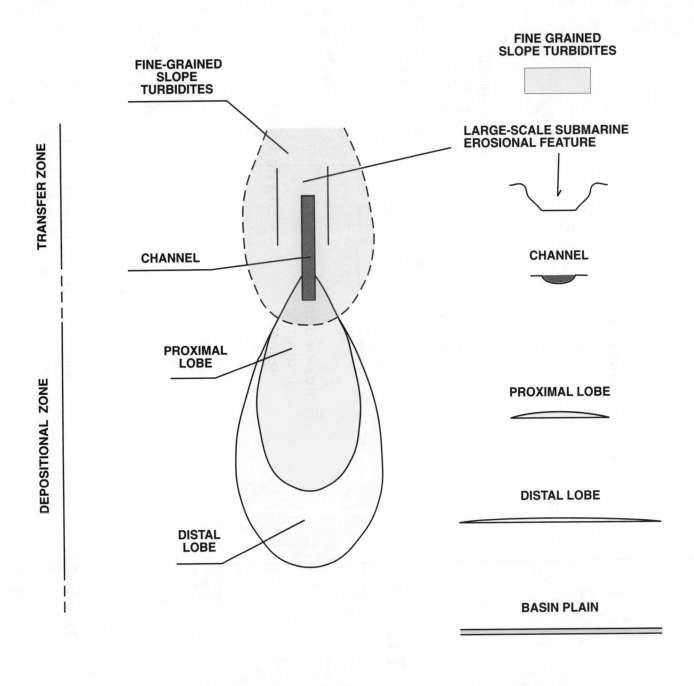

no scale involved

Fig. 18a: Conceptual diagram showing the main elements of a type I depositional system.

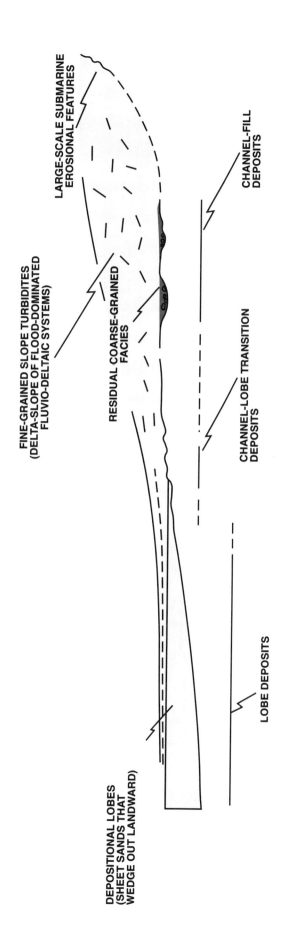

Fig. 18b: Depositional model of turbidite systems and their main component elements (modified from Mutti et al., 1988).

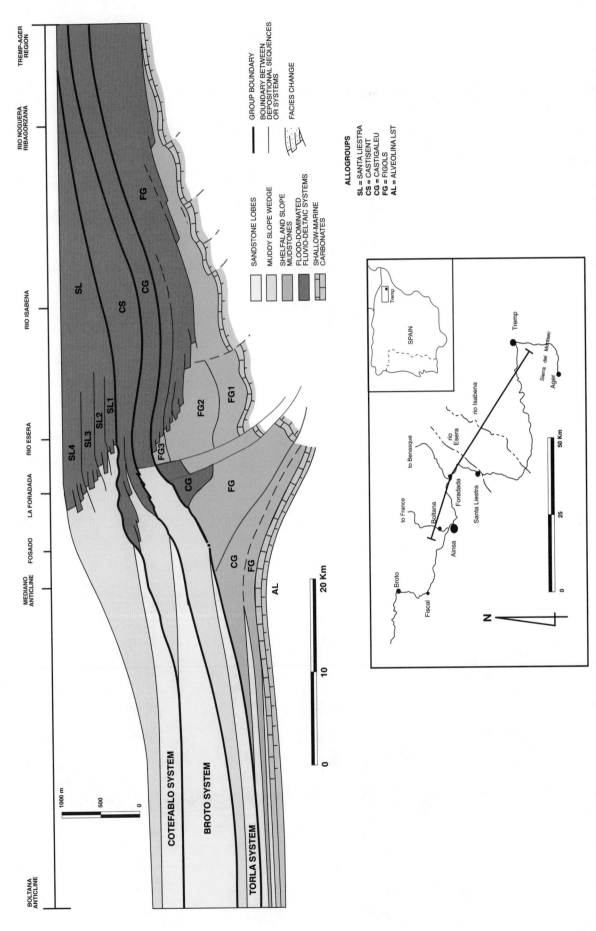

Fig. 19: Schematic cross-section of the eastern and central sectors of the Eocene foreland basin of the south-central Pyrenees, Spain (modified from Mutti et al., 1988).

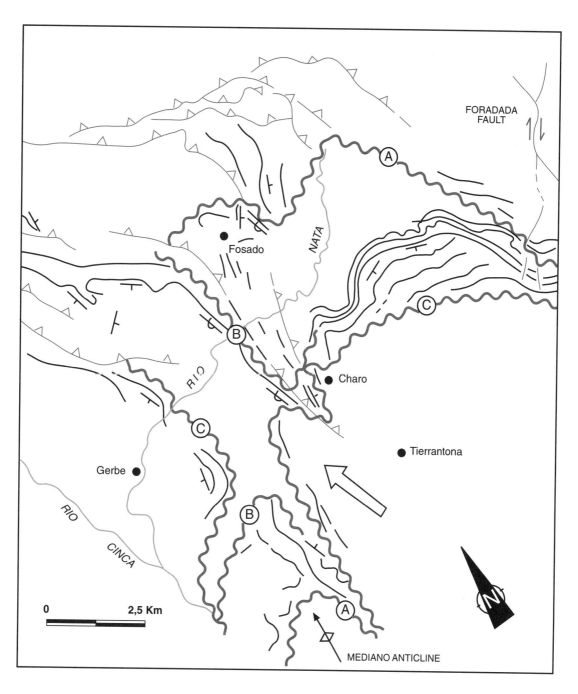

A - Basal unconformity of the Castisent Group
B - Intra-Castisent unconformity
C - Basal unconformity of the Santa Liestra Group

⇨ **General paleocurrent direction**

Fig. 20: Sketch map showing the main submarine erosional unconformities in the marginal portion of the Hecho turbidite basin fill of the south-central Pyrenees (simplified from Mutti et al., 1988).

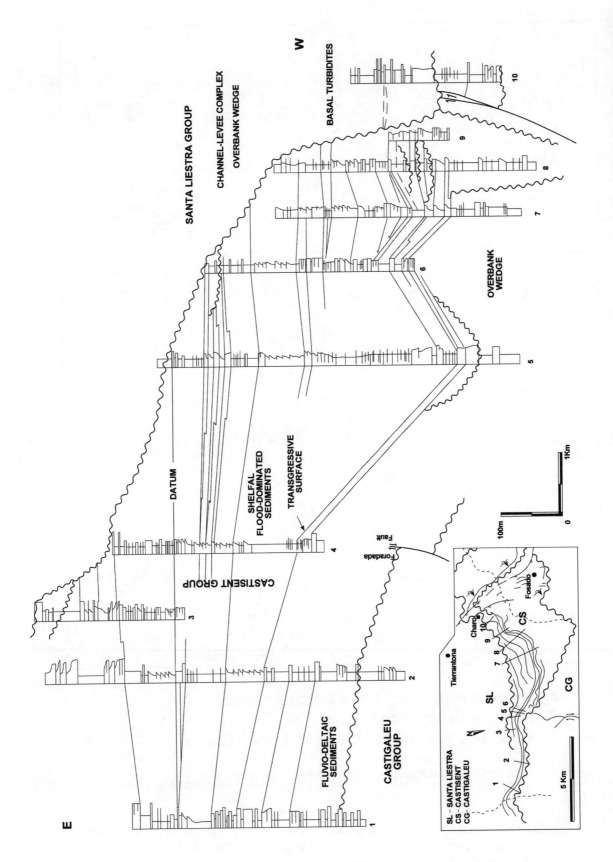

Fig. 21: Example of a high-relief, large-scale submarine erosional feature: Santa Liestra turbidites fill an erosion feature cut into shelfal strata of the Castisent allogroup in the Tierrantona-Charo area (from Mutti et al., 1988).

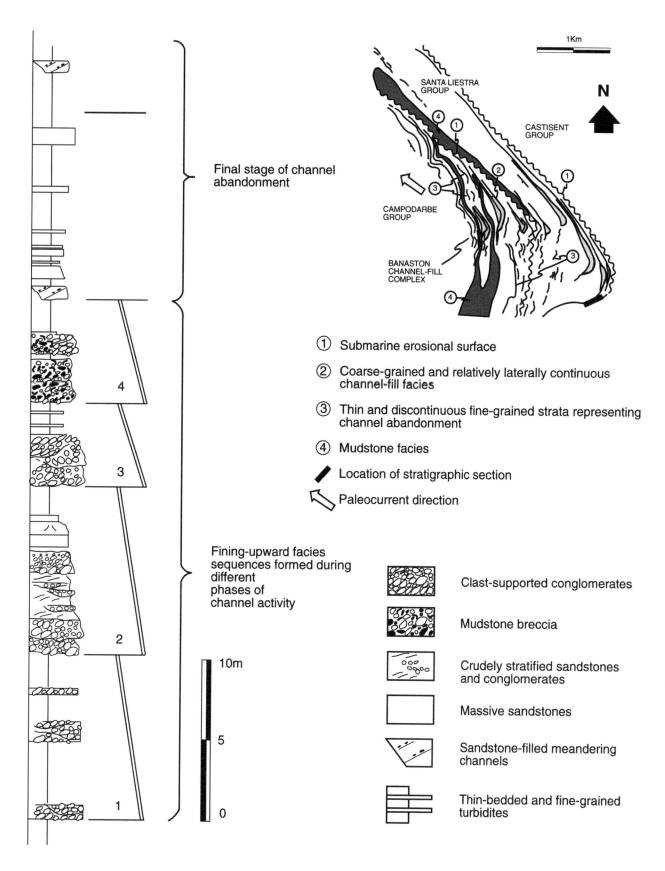

Fig. 22: Overall evolution and facies sequences of the Gerbe channel-fill deposits, Santa Liestra Group (redrawn from Mutti et al., 1985 and Mutti et al., 1988).

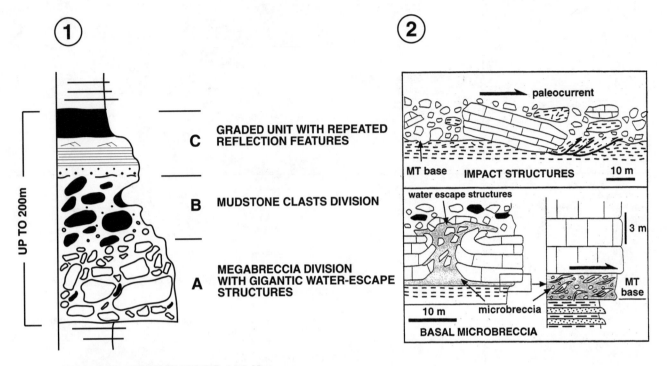
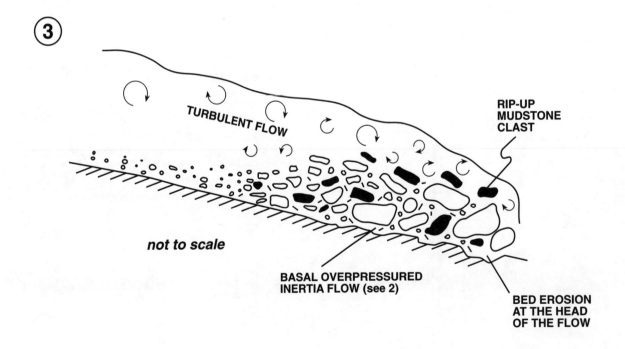

Fig. 23: Internal depositional divisions of a megaturbidite bed (1) and its interpretation (3). Large-scale water escape features are shown in (2), (data from Johns et al., 1981 and Labaume et al., 1987; new interpretation added).

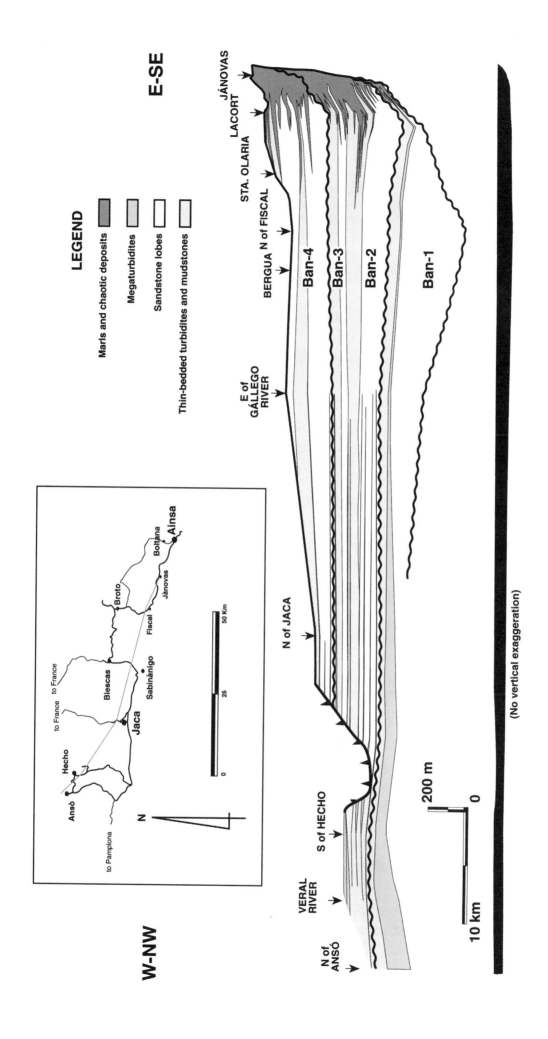

Fig. 24: Cross-section of the Banastòn Group turbidite sandstone lobes in the Jaca Basin, south-central Pyrenees, Spain (modified from Remacha et al., 1998).

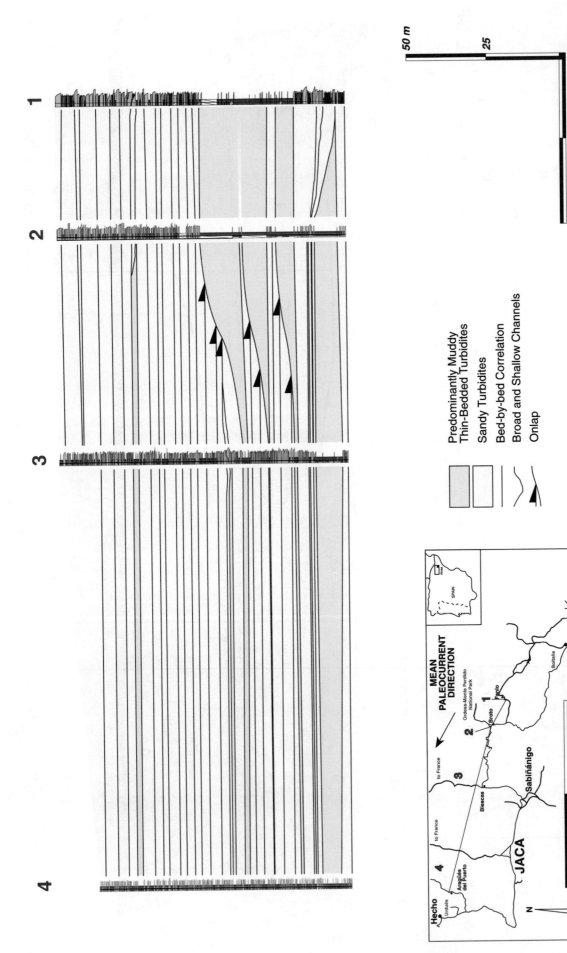

Fig. 25: Detailed stratal correlation patterns of turbidite sandstone lobes from the upper part of the Eocene Broto system south-central Pyrenees, Spain (see cross-section of Fig. 19). Note that the cross section is roughly parallel to paleocurrent direction (from right to left).

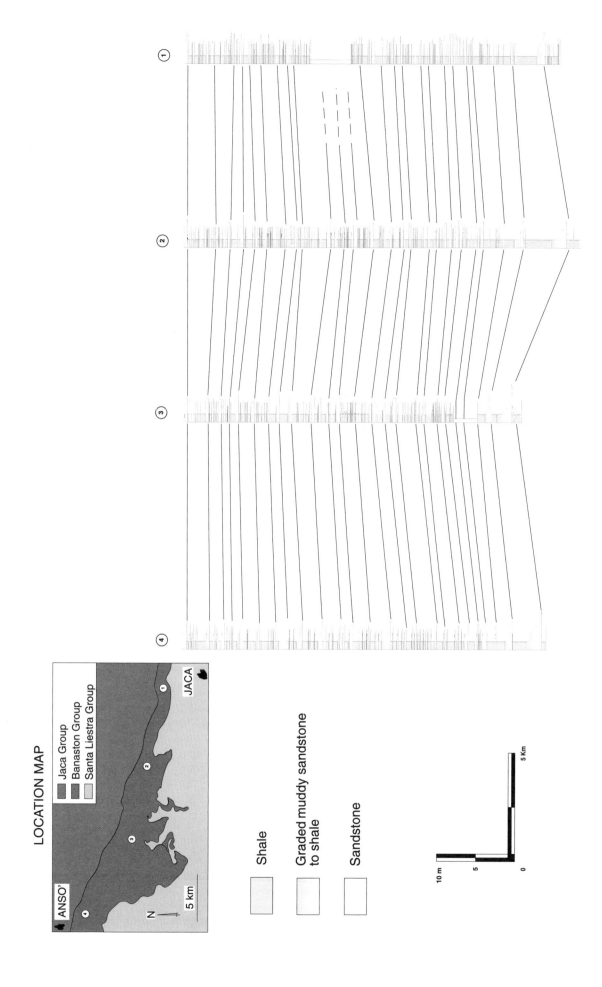

Fig. 26: Detailed bed-by-bed correlation showing the downcurrent transition from distal lobe deposits (right) to basin-plain stata (left), (from Remacha et al., 1998).

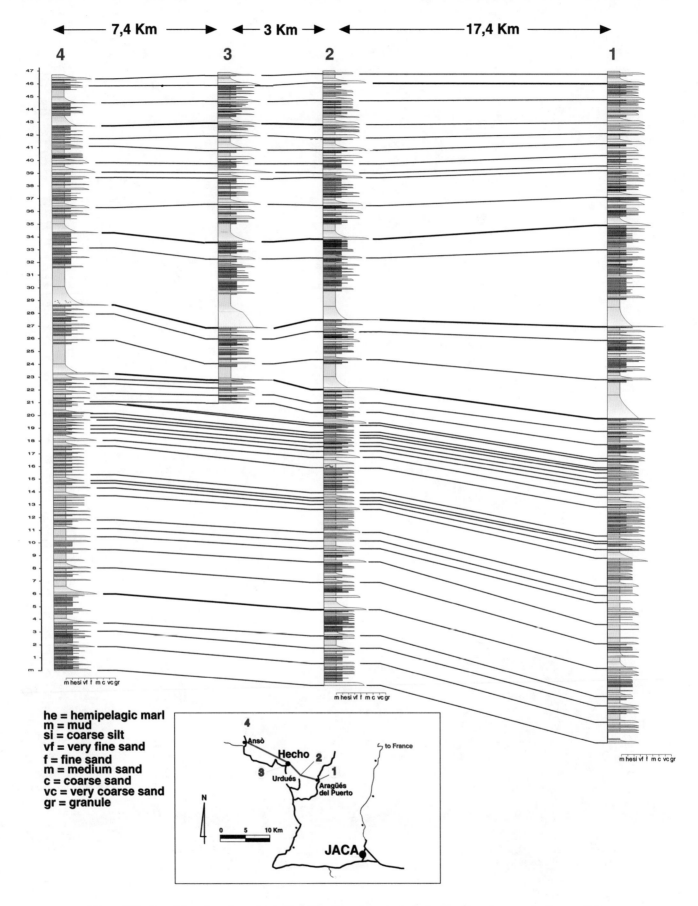

Fig. 27: Bed-by-bed correlation in the basin plain facies association of the Cotefablo system (Santa Liestra Group).

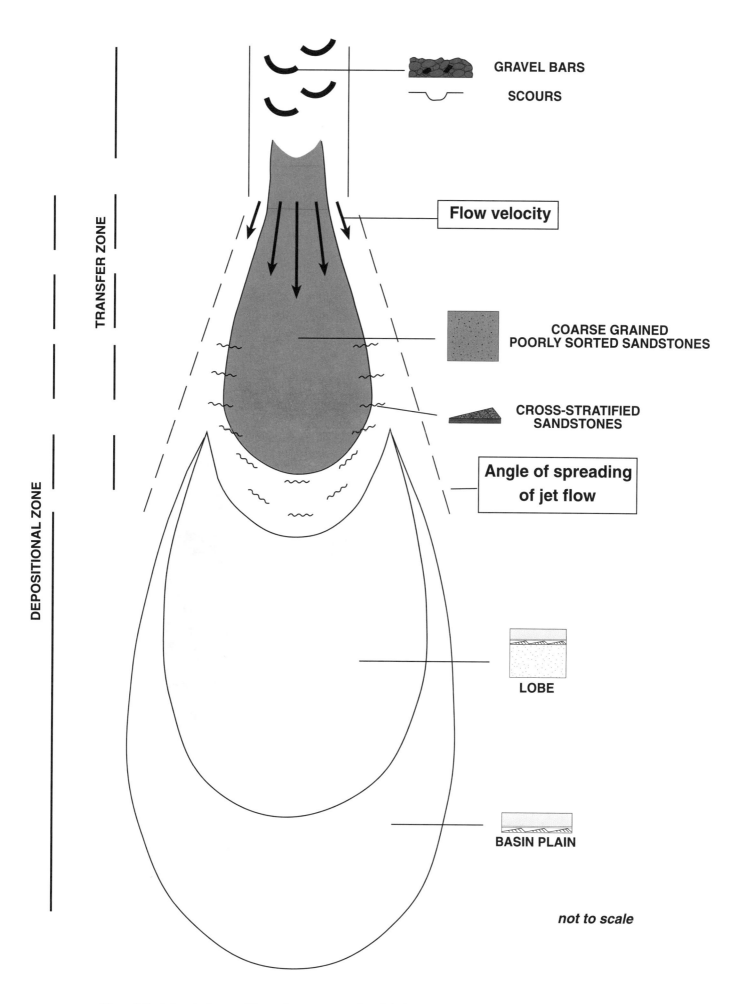

Fig. 28: Ideal depositional pattern of a highly-efficient turbidity current.

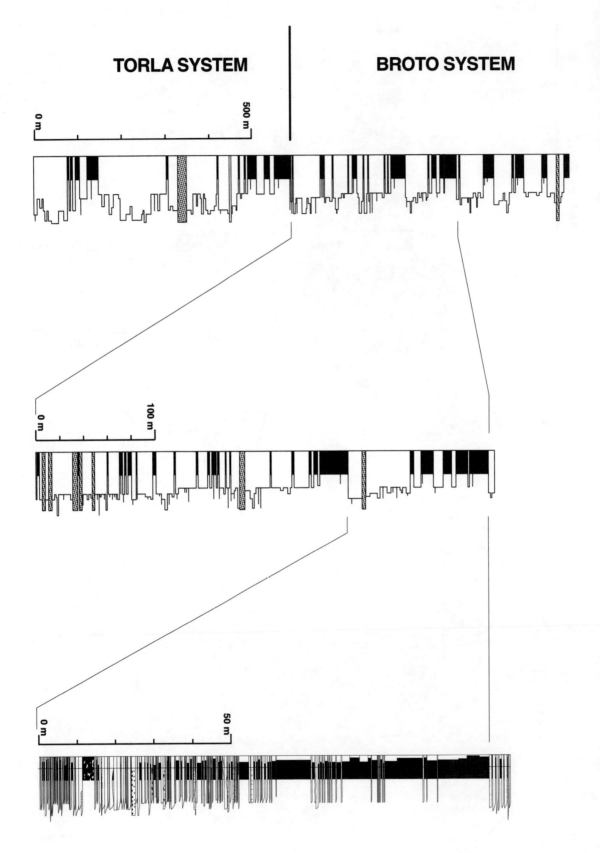

Fig. 29: Hierarchically-ordered cyclic stacking patterns in the Eocene Broto turbidites, south-central Pyrenees, Spain (see cross-section of Fig. 19 for the stratigraphic position of the Torla and Broto systems).

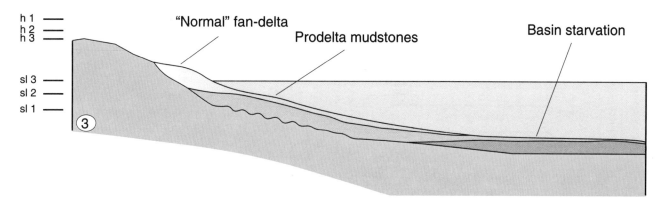

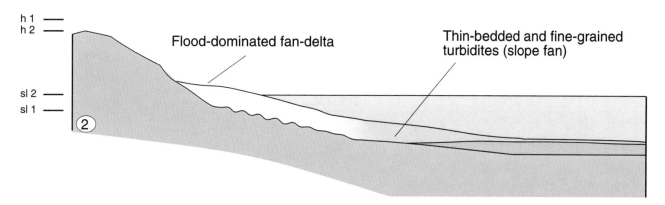

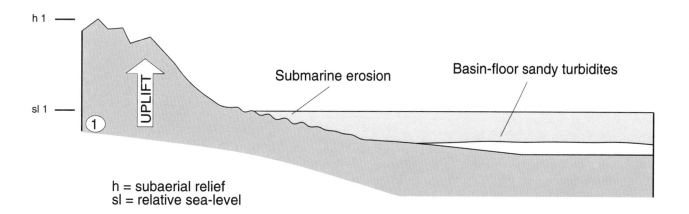

Fig. 30: Diagram showing the kind of unconformity-bounded depositional sequence in turbidite bearing thrust-and-fold basins (from Mutti et al., 1996). Relative sea-level rise is mainly due to subsidence.

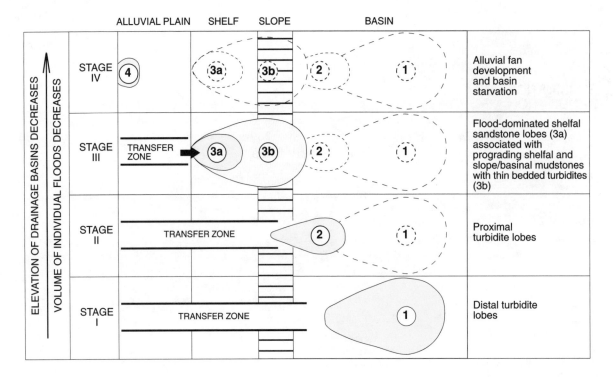

Fig. 31: The different stages of growth of an ideal fluvio-turbidite system formed during an uplift/denudation cycle (from Mutti et al., 1996). Compare with Fig. 30.

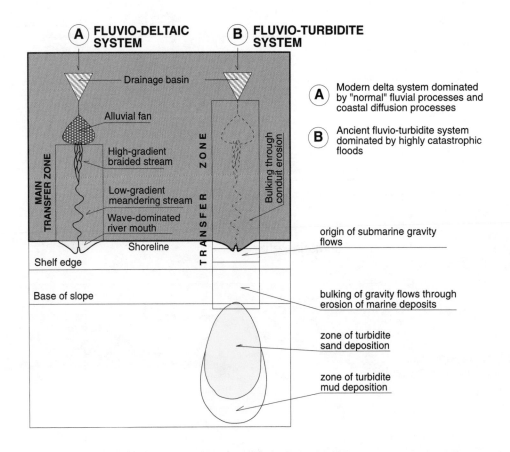

Fig. 32: Diagram showing two different degrees of catastrophism within an ideal fluvial system (from Mutti et al., 1996).

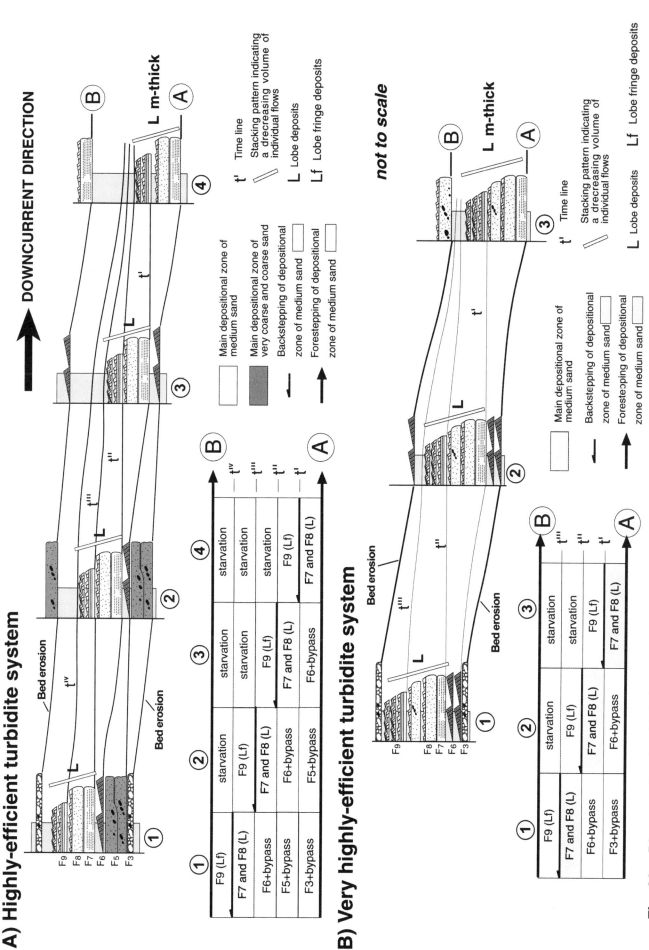

Fig. 33a: Diagram showing high-frequency basinward (forestepping) and landward shifts (backstepping) of the main sandy depositional zones with time in a highly-efficient turbidite system (A) and in a very highly-efficient turbidite system (B).

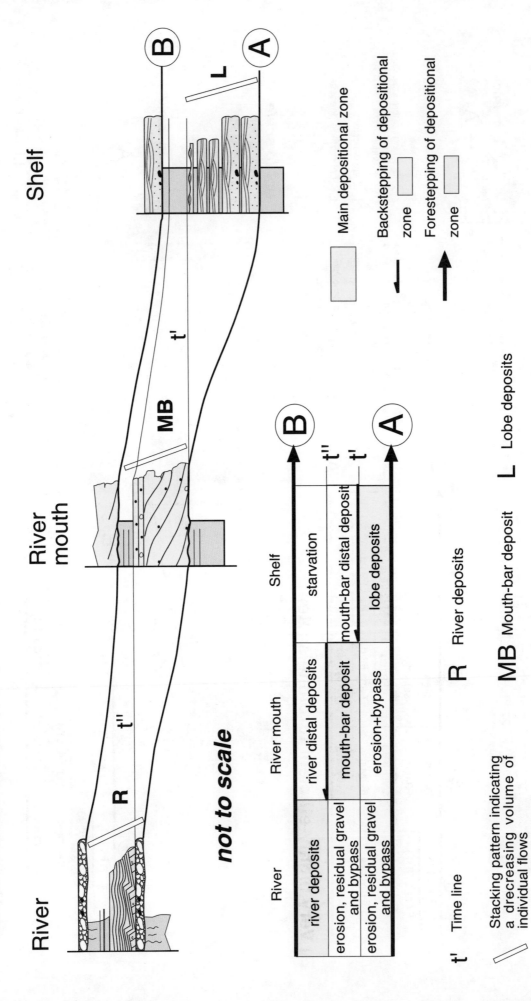

Fig. 33b: Diagram showing high-frequency basinward (forestepping) and landward shifts (backstepping) of the main sandy depositional zones with time in a flood dominated river-delta system (modified from Mutti et al., 1996).